160135

Court Reform on Trial

Court Reform on Trial

on Trial

Why Simple Solutions Fail

A TWENTIETH CENTURY FUND REPORT

Malcolm M. Feeley

Basic Books, Inc., Publishers   *New York*

The Twentieth Century Fund is an independent research foundation that undertakes policy studies of economic, political, and social institutions and issues. The Fund was founded in 1919 and endowed by Edward A. Filene.

Library of Congress Cataloging in Publication Data

Feeley, Malcolm M.
    Court reform on trial.

"A Twentieth Century Fund Report."
Includes index.
    1. Courts—United States.   I.   Twentieth Century Fund.
II.   Title.
KF8719.F43   1983   347.73'1   82–20674
ISBN 0-465-01437-2   (cloth)   347.3071
ISBN 0-465-01438-0   (paper)

For Margaret

Contents

Foreword

LIKE MANY other institutions, the Twentieth Century Fund has been concerned about the explosive rise in crime. With our limited resources, we could not undertake major examinations of the various components of the criminal justice "system" that, in any case, were being studied by both private and public agencies. Instead, we decided to establish a series of independent Fund Task Forces to investigate discrete problems affecting criminal justice and to make recommendations for treating them. Thus, our Task Forces were examining the role of the Law Enforcement Assistance Administration and sentencing policy, one report concentrating on adult criminals, the other on juveniles.

Our focus on sentencing policy naturally led to a consideration of the place of the courts, which were widely reported to be in a state of crisis. The courts, of course, were a critical component of the so-called system and had been subject to many efforts designed to reform their alleged failings. And just as the Fund's research staff began to search for scholars familiar with the courts and their workings, we received an inquiry from Malcolm M. Feeley, who expressed an interest in a critical appraisal of court reform.

Foreword

Mr. Feeley's credentials and his proposal impressed the staff and the Fund's Board of Trustees, which gave its approval to his project. An objective and thoughtful scholar, he wanted to evaluate a number of court reforms, explaining how they had developed, what claims were made for them, what differences they made once put into effect, and what should be done now. In the course of his research he came up with many useful insights and findings, among them that the changes—the reforms—imposed upon a system are, in the course of events, altered and shaped by the different participants in the system as well as by the system itself.

Perhaps his most significant finding is that the crisis in the courts has been exaggerated and distorted. This does not mean that Mr. Feeley takes a complacent view of things. His carefully detailed analysis of major court reforms—pretrial diversion, speedy trial rules, sentence reform, and bail reform—makes clear that he has an incisive understanding of the problems they were supposed to alleviate and the difficulties they encountered. His is an instructive and cautionary tale of what reforms can and cannot do, one that has special pertinence to the courts but is no less applicable to other long-established institutions.

The Fund is grateful to Malcolm Feeley for his commitment to his study and to the meticulous and evenhanded way in which he carried it out. It should provoke public debate. It should also contribute to better—if perhaps more modest—policy making.

M. J. Rossant
DIRECTOR
The Twentieth Century Fund
July 1982

Preface

POLITICIANS lash out at the criminal courts for their failure to protect the community. The media describe them as assembly lines. Scholars characterize court officials as self-serving opportunists. On occasion, judges open up and reveal troubled hearts beneath their somber robes. The consensus: there is a crisis in the courts.

This crisis, according to some observers, stems from the rapid rise in crime and the staggering case loads created by that rise. Others argue that it is caused by a concern with due process run amok. Still others point to uncaring officials.

Clearly, the courts face serious problems. Crime rates have mushroomed; arrestees languish in overcrowded jails or are freed to walk the streets for weeks, months, or years before their cases are decided; recidivism is rampant; witnesses and victims meet with such indifference that they begin to regard the court process as another form of victimization; and judges cede much of their authority to prosecutors, who barter it away for guilty pleas. Courtrooms and jails dramatize the sharp divisions of class and race in American society. Concern about crime and the problems of the courts is very real, neither figments of the imagination nor overreactions to isolated events.

But acknowledging that the courts have a great many problems does not mean that they are on the brink of collapse. Even if the problems they face are unprecedented, the problems must be understood in context and in perspective. Of course, every situation is unique; no two events are ever identical. But if this view is carried to the extreme, if we accept the idea that the problems of the courts are due to new and historically unique events and that it is more misleading than instructive to analyze past practices as a guide for coping with future problems, there will be nothing to guide our judgments and our inquiry.

The facts, however, do not lead to the conclusion that the problems facing the courts are unique or even that there is a crisis in the courts. They do not even lead to the conclusion that the courts are facing unprecedented problems.

The exaggerated assertions about the problems faced by the courts, the historical perspective that informs (or, more properly, fails to inform) so much analysis, and the easy—and often wrong—answers implied in so many crisis-generated discussions are, however, in need of examination. Courts are not what they appear to be. The nature of their problems is not always evident. The law as written does not capture the real structure and operations of the courts. One of the central problems of the courts is that there is no agreement on what constitutes acceptable practice and hence no agreement on what improvements should be made. Practices that are regarded by some as signs of decline may, when seen through someone else's eyes, be seen as strengths.

The discovery of social problems reveals the expectations of the viewer. A crisis may be the consequence of a dramatic change in events, but it also may be manufactured, a function of increased attention by the media and altered expectations. Both types of crisis are real; both have consequences. It is important to distinguish the two types of events—dramatic turns of history and shifts in expectations—because the strategies needed for coping with each will be different.

The primary problems of the courts are not the result of external calamities, such as increased crime rates and case loads; rather they are due to changes brought about by raised standards and increased attention. By characterizing the problems of the courts as a crisis, politicians, the press, the scholarly community, and the courts themselves have misdiagnosed the real problems they are facing, fostered unrealistic expectations, and promoted bold but often empty solutions that are guaranteed to bring about disillusionment and disappointment even in the face of significant improvements.

Like most public service institutions, courts are complex organizations. Informal practices and procedures are not idiosyncratic accidents but are usually the result of perceived necessities. Focusing on the shortcomings of a single practice without placing it in historical and functional context usually leads to gross distortion and exaggeration. Efforts to eliminate such problems without altering incentives may result in their reappearance in another and perhaps more serious form. The courts are an institution whose powers are extremely limited; yet they are frequently called upon to perform Herculean tasks.

It is naïve to expect an institution into whose clutches only a handful of criminal offenders fall to have a measurable impact on either those who come within its grasp or those who successfully elude its grasp. What the family, community, workplace, school, and religious institutions have failed to achieve cannot be accomplished by a brief encounter in the courts, however speedy or deliberate, lenient or harsh.

To say that things have gotten better in the courts while pointing out their limited efficacy is *not* to endorse the status quo. Rather it is a plea to establish realistic expectations, to learn to live with inevitable tensions, and to acknowledge that fundamental social changes are outside the courts' capacities.

This book examines the systemwide impact of several crisis-spawned innovations. My thesis is that, because our understanding of the courts is flawed and our expectations about what the courts can do are unrealistic, many innovations fail. Some fail because we try to impose bureaucratic structures on a protean if imperfect adversary process. Some fail because we mistake discretion for arbitrariness. Some are misdirected because we focus on isolated horror stories. Others respond to the symbols of legal formalism and ignore actual practices. Some strive to extend the courts beyond their capacities.

Conversely, many concerned observers have unrealistic expectations about what courts can do, while others fail to appreciate that some of their concern stems from increased standards and expectations; as a result, they underestimate the significance of changes that have taken place.

To explore this thesis, four quite different innovations are examined: bail reform, pretrial diversion, sentence reform, and speedy trial rules. Each has been promoted as a way to achieve an important objective and has been replicated in a number of jurisdictions. While each has its own set of vigorous critics, each is generally considered to be a serious and important enterprise. And because it would serve little purpose to show that ill-conceived and poorly supported projects have had problems, my investigation focuses on the more promising examples of these reforms. Yet, as I will argue, each has to varying degrees and in different ways been based upon erroneous assumptions and exaggerated expectations and, as a consequence, is something of a failure. My objective is to explain those failures and, in so doing, explore the nature of criminal courts.

Acknowledgments

IN WRITING *Court Reform on Trial,* I owe a singular debt to the staff and director of the Twentieth Century Fund. The opportunity to pursue this topic came along at just the right time, and the Fund's generous support allowed me time to develop my ideas carefully. Bill Diaz initially helped me clarify my ideas, and later Carol Barker and Gary Nickerson were valuable critics of successive drafts. Beverly Goldberg and Joan Schwartz are my ideals of good editors: insightful, rigorous, and right. I appreciate the Fund's willingness to provide such talent.

The perspective developed in this book owes a great deal to a large number of people on whom I have inflicted my ideas. The late Hal Chase first sparked my interest in criminal court matters over fifteen years ago and continued to serve as a periodic sounding board for me until his death. Dan Freed initially kindled my interest in most of the problems examined here. Roger Hanson, Sam Krislov, Barry Mahoney, and Paul Nejelski all have listened to me discourse on this project more than they probably cared to, and Roger gave me valuable comments on an early draft.

Others with whom I have discussed at length issues in this

Acknowledgments

project or whose writings have been especially influential in shaping my thinking include Murray Edelman, Linda Fleming, Wallace Loh, Jon Silbert, Michael Smith, Stan Wheeler, Herman Goldstein, Ray Nimmer, Frank Zimring, Norval Morris, David Rothman, Al Blumstein, and Bill Nagel.

Court Reform on Trial examines a number of quite different topics, and because of this I rely heavily on the publications and advice of others who have long experience working in one of the areas. Lee Friedman, Michael McConville, Alan Henry, and the late Woody Dill were helpful in my investigation of pretrial release, and the last three carefully read one or more drafts of the chapter on bail reform. Dan Ryan, Floyd Feeney, Madeleine Crohn, and Sally Baker were helpful in informing me about pretrial diversion, and Sally saved me from many errors with her close reading of a draft chapter on that topic. Frank Remington, Russ Wheeler, and Dan Freed were especially helpful in providing me with background information on speedy trial legislation, and Tom Church and Ken Mann provided valuable comments on my chapter on speedy trials. I drew on a great many sources and people in my examination of the impact of sentence reform: Shelly Messinger introduced me to the new California sentencing law, directed me to additional sources, and periodically provided me with reports on recent developments; Jay Casper and Keith Hawkins carefully read my chapter on sentence reform and suggested helpful improvements.

Lawrence Friedman and James Q. Wilson read my last version of the manuscript and helped me tighten it up. I am also indebted to several anonymous readers selected by the Twentieth Century Fund. Their detailed and insightful comments on a draft of the manuscript helped immeasurably to improve the final product.

Dozens of public officials gave generously of their time to

Acknowledgments

respond to my queries as I visited courts from Brooklyn to Oakland, fired off batches of letters, placed innumerable telephone calls, and waded through stacks of reports and documents. Although I raised difficult questions and was obviously skeptical of many claimed accomplishments, I generally received candid and frank answers. For this I am deeply appreciative.

This project benefited from two talented research assistants. Randy Levinson helped me as I got launched into my research, and Mark Lazerson put his legal training and writing ability to good use. Mark assumed considerable responsibility in developing the chapters on sentencing and speedy trials and must share credit with me for whatever virtues they possess.

I have benefited from the staffs and facilities of several institutions. My thinking was clarified by presentations at the University of Toronto, the University of Washington, and Birmingham University, and in discussions with colleagues and students at the University of Wisconsin. The staffs of the Yale Law School Library, the Federal Judicial Center, the National Center for State Courts, the Institute for Advanced Legal Studies, and the ABA Action Commission to Reduce Court Costs and Delay all provided me with important information or granted me the use of their facilities at various stages of this project. In particular, I must thank Barbara Meyer and her colleagues at the Criminal Justice Reference and Information Center at the University of Wisconsin. Buried away in that basement is an outstanding criminal justice collection as well as a staff of the most helpful librarians I have ever met.

Court Reform on Trial could not have been completed without an understanding typist, and I had the best. Lyn Reigstad could decipher my handwriting even when I could not and provided clean copy for successive drafts at a moment's notice.

Finally, I must acknowledge my wife, Margaret. She helped in more ways than I can enumerate.

Part 1

Introduction

Chapter 1

THE COURTS AND CHANGE

CONSIDER this chilling composite.

A twenty-two-year-old male is arrested for burglary and assault with a deadly weapon, both felony offenses carrying maximum penalties of ten or more years. At arraignment the judge notes that the accused has a prior record and faces serious charges involving a gun. He sets bail at ten thousand dollars. Unable to afford a bondsman's fee, the accused is remanded to the county jail. Two months later the judge reduces the bond to twenty-five hundred dollars, and relatives of the accused scrape together enough money to secure his release. Three months and six court appearances later, he pleads guilty to a single count of criminal trespass, a misdemeanor, and receives a five-month sentence with two months credited for time served in pretrial custody.

Everyone involved agrees that this case constitutes a problem. The arresting officer, who strongly suspects this offender of a string of burglaries, will point to the reduced bail and light sentence as evidence that the courts do not care. The public defender who is handling the case will point out that his client, presumed to be innocent until proven guilty, spent two months in jail solely because he is poor. An attorney familiar with the case might privately admit that, if his colleague had reviewed the police report carefully at arraignment, he could have challenged the assault (gun) charge at the outset, possibly obtained the immediate release of his client, and probably avoided the jail sentence altogether. The defendant's family is distraught because of the financial hardship the case has brought on them. The prosecutor—the third assigned to the case—is frustrated because she could not locate one key witness and another proved unreliable. The judge is irritated because the case appeared on his calendar eight times before it was finally resolved. A social worker attached to the court is concerned because the defendant has an alcohol and drug problem that no one is doing anything about. After several trips to court, the proprietor of the burglarized store feels twice victimized—not only did he lose money during the robbery, but he has now lost money every time he left his store in order to appear in court. The one available witness, a passerby who claims to have recognized the intruder and alleges that he hit her with the barrel of a gun before he fled, is indignant at the lack of respect accorded her by the prosecutor. For his part, the defendant claims that, finding the door to the drugstore open, he entered to see if anything was wrong. In the aftermath of the case, the defendant might cynically conclude that he got off easy or that his attorney—a public defender whom he did not know, did not pay, and rarely saw—had sold him down the river.

Even if he could have heard the exchanges above the noise, a first-time observer in the courtroom would not have been able to understand what was going on. But, noting that the accused was black and that all those in a position to affect his fate were white, he might have drawn a conclusion of racial discrimination. A writer covering the courts during this period might write an article using this case as an example of the bargain-basement inducements used by the courts in order to cope

The Courts and Change

with overcrowding, both of the courts and the jails. Sometime later, at a meeting of the county board of commissioners, a budget analyst might cite this case in a report on the fiscal implications of judicial leniency—county, not state, governments typically bear the cost of maintaining inmates sentenced to jail for periods of one year or less. Including this case in a sample that produces a statistically generated profile of burglars, a criminologist might estimate that this type of offender has probably committed a dozen or more burglaries and is likely to be rearrested within the next two years.

No one is satisfied. All of them think they have a solution: pretrial detention, release on recognizance, full enforcement, elimination of overcharging, expansion of police powers to interrogate, better attorneys, fixed sentences, improved prison financing, rehabilitation programs, more coordination, victims' rights, attacking root causes.

Some of these solutions, of course, cancel one another out. Others are beyond the reach of the courts. Many are based upon incomplete or inaccurate information.

Let us again consider the case: although the accused was charged with possession of a deadly weapon, the police were not able to produce a gun. The only *evidence* about the gun is the statement by the woman who claimed to have seen and been hit by it when she encountered the accused in the doorway to the drugstore; the druggist did not report seeing a gun, and although the accused matches the druggist's general description of the burglar, the druggist is very hesitant when making a positive identification. The police report also states that the woman did not appear to be seriously injured, that she refused offers of medical assistance, and that she "seemed intoxicated." She also failed to keep some of her appointments with the prosecutor, and a number of details changed each time she recounted the incident. And although the proprietor of the drugstore reported the loss of several cameras and transistor radios, in addition to cash, and the accused was quickly apprehended based on the key witness's identification, none of the items listed was ever recovered, and the money involved was not large or identifiable. Prosecutors privately acknowledge that there is an incentive for property crime victims to exaggerate losses for insurance and tax purposes.

There were various reasons for continuances: a probable cause hearing, a conflict in the defense attorney's schedule, an apparent malfunction in the computerized calendar system, a sudden resignation in the prosecutor's office, then a vacation, and an effort to locate another witness who had moved out of town.

Was the court lenient? Although the defendant had a history of prior arrests, he had only two prior convictions, both on breach of peace, and had never done time. Court records reveal that five months is above average for trespass cases. Harsh? Frustrated by the inefficacy of probation and by reports of the failures of drug and alcohol treatment programs, the judge feels he has no option other than to put the offender behind bars. But aware that violence and sexual abuse are commonplace in state prisons and impressed that the offender's family had stood beside him, the judge hopes that he will be safer in the local jail and that he will be better able to maintain ties with his family. Still, the judge, ambivalent about the sentence, expresses hope that the offender may be eligible for daytime work release after a month or so, unaware that this program has been eliminated in recent budget cuts.

Although a composite, the process just described illustrates the range of issues raised every day in courthouses across the country. As troublesome as it is, the above picture is an improvement. Twenty-five years ago, the accused would probably

The Courts and Change

have remained in pretrial custody until conviction. He might have been subjected to sustained interrogation by the police and without benefit of counsel might have pleaded guilty to the original charges and other burglaries as well. Or had he been able to scrape together money for an attorney, the attorney might have traded on his political connections and perhaps some cash to pave the way for swift release and dismissal of charges. Not too long ago, the option of even considering alternative treatment programs would not have been available. And until a few years ago, if this incident had occurred in certain regions of the country, and if the questionable witness had been white, the prosecutor might not have challenged her credibility.

Despite frustrations, the seemingly intractable nature of the problems facing the courts, and the hydraulic quality of the courts, changes have occurred. Whether these changes constitute an improvement is, of course, a matter of judgment. And as we see, one of the problems facing the courts is that they are held accountable to a bewildering array of standards by people with quite distinct views. While I am explicit about my own point of view, the purpose of this book is more generally to examine attempts at planned change in order to illuminate the nature of courts. The terms *reform* and *innovation* are used to convey the perspectives of their proponents and not necessarily my own views.

In the past two decades, the four major innovations I examine (bail reform, pretrial diversion, sentence reform, and speedy trial rules), have been tried repeatedly in order to overcome some of the problems alluded to above. Each has been planned with an important objective in mind. Each has been replicated in a number of jurisdictions. Each is considered a serious and important enterprise. Each needs to be examined in detail, as does the process of change itself: What is necessary for successful change, and at what critical points can a particular innovation succeed or fail?

If we understand the structure and the functions of the courts, and if we are able to view them in historical perspective, we can see that there has been change over time. Now it is important to determine whether this change is attributable to the innovations that will be examined or whether the change has taken place in spite of or concurrent with them.

The four reform efforts that must be analyzed more fully are:

Bail Reform. Pretrial release agencies, as they are generically known, have been set up to supplement or replace traditional money bail by offering nonmonetary release on recognizance. They rely upon the "expert" opinion of disinterested third parties who collect and review information on the accused's ties to the community and make an informed recommendation to the judge about pretrial release. The Vera Institute of Justice, in New York City, is the recognized pioneer in this type of bail reform.

Pretrial Diversion. This refers to formal programs in which defendants are enrolled, thus eliminating the need for adjudication by the courts. Typically, these programs offer short-term counseling, job training, or placement services in lieu of prosecution. (The Manhattan Court Employment Project has been the model for such programs throughout the country.)

Mandatory Minimum and Determinate Sentencing. These are efforts to curb judicial discretion in the sentencing process. Although they rest on quite different philosophies, both are designed to impose constraints upon judges at sentencing. Four well-known sentence reforms are the Rockefeller Drug Laws in New York, the Bartley-Fox Gun Law in Massachusetts, the

The Courts and Change

Michigan Mandatory Minimum Firearms Sentencing Law, and the California Determinate Sentencing Act.

Speedy Trial Rules. Speedy trial rules attempt to make practical the Sixth Amendment guarantee of the right to speedy trial. While these rules take many different forms (for example, court decisions, court rules, legislation), their common purpose is to establish time limits within which charges against the accused must be readied for trial. Many speedy trial rules provide for dismissal of charges if time limits are exceeded. The Federal Speedy Trial Act of 1974 is important both in its history and its implications.

Fragmentation: The Various Meanings and Functions of Courts

The process upon which certain important reforms have been grafted has been shown at work; the structure underlying that process must also be examined.

The criminal court is a political agency, a public service institution that operates within the constraints of scarce resources and faces conflicting and competing demands from a multiplicity of publics and the organizational exigencies of its own personnel. Perhaps its most visible quality is fragmentation: it is fragmented in its organization, its operations, and its goals. This is not an aberrational feature of the court, which traditionally has been a flexible and responsive institution. In order to fully appreciate its function, its flexibility must be accepted along with its lack of efficient organization and coherence. The courts should not be seen merely as bureaucratic organizations committed to clear and well-defined purposes—they should be understood as arenas in which a range of competing and conflicting interests collide and vie for attention. Fragmentation, with its accompanying dislocation and disagreements, is natural to them. The courts are run on a system of interaction whose efficiency and end product are more akin to Adam Smith's notion of unplanned, unconscious coordination in the pursuit of self-interest than to any theory of rational organization.

Courts and Games. It is instructive to look at fragmentation in the courts in terms of a multiplicity of games. Judges, prosecutors, defense attorneys, defendants, clerks, police officers, bailiffs, sheriffs, bondsmen, witnesses, and all the others whose activities take them into the courthouse pursue distinctly different interests and purposes and may understand their participation in the process in entirely different ways.

Each of these players is participating in a different game, or in several different games. The games may or may not be centered in the court and may or may not directly involve other participants in the criminal process. It is also important to understand that the courtroom is simply one of several arenas in which these games are played; actions outside the court also affect the way the criminal process is structured.

For instance, the police are engaged in a public order game, responsive to citizen complaints about domestic violence, public disturbances, and personal safety. One result: they often make arrests to keep the peace, without any intention of initiating successful prosecution. Or they overcharge to compensate for other offenses they cannot prove.

Prosecutors are involved in public order and civic virtue games. "Bad" people are vigorously pursued; when serious charges cannot be proven, lesser, more easily proven charges may be invoked, so that the bad people do not escape official sanction altogether.

Defense attorneys have their own variation of this game.

The Courts and Change

Criminal procedure, rules of evidence, and threat of trial are not only channels within which the search for truth is conducted—they are also used instrumentally in the effort to obtain the most desirable outcome.

The *court* in this perspective becomes a convenient shorthand term for a collection of individuals who, while sharing common concerns and interests and knowing that they must work together to process criminal defendants, are also engaged in a variety of other enterprises not formally acknowledged in the law. It is the combination of the two sets of functions that gives meaning to their actions and shapes their activities. From this broader and more functional perspective, court behavior, which is often characterized as confusing, is more understandable. Indeed, it becomes normal.

THEORETICAL BASES OF FRAGMENTATION

Fragmentation in the courts is reinforced by the theory behind the American criminal justice system, which has three basic components: the adversary process, due process, and professionalism.

The Adversary Process. In his book *Courts on Trial,* Jerome Frank characterized the adversary system as embodying a "fight theory of justice."[1] He traced its origins to actual physical combat, in which participants in a dispute sought to ascertain truth by means of a duel or contest. From this ordeal by battle, there slowly emerged ordeal by words. Although the weapons of combat have changed, the underlying theory has not. It remains: truth is most likely to emerge through active combat between partisans, through attack and counterattack. The intensity of self-interest, so holds the theory, maximizes the likelihood that truth will emerge. It is, in the words of Richard Posner, as if the court were "in the position of a consumer forced to decide between the similar goods of two fiercely determined salesmen,"[2] each pointing out the benefits of his own product and the deficiencies of his competitor's.

The analogy to the competitive market is instructive. Like the theory of the free market, the theory of the adversary process rests on a belief in the capacities, initiatives, and self-interest of participants. The primary role of neutral third parties in both the legal and the economic sphere is to see that the adversaries pursuing their own distinct and clashing interests do so according to the "rules of the game."

Contrast this view with the argument that truth is best pursued and most likely to be obtained by someone who has no stake in the outcome, a disinterested inquisitor who investigates all sides of an issue and makes judgments. One might think that the judge—the disinterested party who must render a verdict—would assume this role, but he does not. The judge, in the Anglo-American tradition, does not set the charges, has no independent investigative staff, and possesses little authority to seek out witnesses. Rather, his role is to sit back and umpire the process as it unfolds before him. What the adversaries do not present, the judge does not consider.* Indeed, if they arrange to settle the dispute prior to appearing before him, in all but rare instances the judge does little more than ratify their decision. Many judges even cede their sentencing authority, content to prescribe the term proposed by the prosecutor.

In the United States, there is no ministry of justice, no criminal justice czar, no one to see that everyone works together to pursue common objectives. Rather, there are distinct of-

---

*Judge John Sirica was widely praised by the public for his tenacious efforts to uncover the "truth" in his handling of the lesser offenders in the Watergate scandal. But he was severely, although quietly, criticized by his judicial brethren who felt that his tactics—conditionally imposing long sentences on those lesser offenders in an effort to induce them to reveal incriminating information about the "higher-ups"—severely strained the judicial role. Judge Sirica's response to his critics was that he used as best he could what limited resources he had at his disposal. This exception is, I think, ample demonstration of my point.

fices—police, prosecutors, defense attorneys, judges—drawn apart still further by the doctrine of the separation of powers and the theory of the adversary process. Like market theory, adversary theory relies heavily on the belief that from the clash of partisan interests, the most satisfactory results will emerge.[3]

Due Process. One of the animating features of due process is fear of authority, a concern with the potential for abuse of power by the state. It is for this reason that functions are separated, authority fragmented, and power circumscribed. Hence, even the prosecutor, who is an official of the state, is regarded in many respects as if he were a private individual, someone whose power and capabilities are roughly equivalent to those of the defense.[4] The power of the state is diminished still further by the insulation of the judiciary. Judges comprise a distinct and independent branch of government, presumably far removed from the pressures of the executive enforcement officials.

The rules of criminal procedure—due process—force the state, to borrow Herbert Packer's phrase, to run through an obstacle course before someone can be judged guilty. In order for evidence to be admissible in court, it must be obtained without violating strict rules. The accused cannot be compelled to testify against himself. Changes must be clearly specified. The accused has the opportunity to confront witnesses and offer defenses. These rules are designed to do more than accurately determine whether the accused did or did not actually engage in the crime in question; they are designed in part to restrain officials, regardless of guilt.

The detailed formalism of criminal procedure facilitates fragmentation; so also does its open-ended discretion. Despite the popular conception of courts as adhering to abstract rules at the expense of common sense, and despite the rigid formality of many stages of the legal process, criminal procedure permits vast discretion at each of several critical stages. Prosecutors have virtually unlimited and unreviewable discretion in setting charges and in deciding whether or not to prosecute at all. A substantial portion of convictable cases do not go forward for no other reason than that the prosecutor feels the interests of justice are not best served by prosecution. Often, this occurs because the accused has agreed to make restitution, or to join the army, or the complaining witness decides to make an appeal on the behalf of the accused. It is but a small step from this traditional form of discretionary practice to plea bargaining, a process of dropping or reducing some charges in exchange for a plea of guilty and the assurance of likely sentence.

A cornerstone of due process is the right to trial. Despite this, there is no prohibition against waiving the right. Indeed, the primary responsibility for invoking opportunities and rights is in the hands of the accused himself. This key and critical event in the obstacle course is only optional. So, too, are other steps in the criminal process: the right to rely on counsel, the right to challenge witnesses, and the right to confront accusers. The optional nature of this process contrasts sharply with the theory of inquisitorial process of many European countries, among them France, Germany, and the Scandinavian countries. In these countries the law assigns responsibility for developing cases to an independent investigator who is to pursue his task of making a case against the accused regardless of the wishes of the accused and, often, of the victim as well. Ironically, in theory, the American adversary system rests upon the presumption of innocence and requires the state to prove guilt; yet in practice American courts rely almost entirely on the guilty plea, which eliminates the state's need to *formally* prove its case. In contrast, European inquisitorial systems—long popularly associated with the abuses of coerced confessions and the like—rely extensively on the report of an investigator charged with the task of marshaling *all relevant material on the case* and depend less on self-conviction through pleas of guilty.[5]

Sentencing is another critical stage at which American criminal procedure allows considerable freedom. By law and custom in the United States, sentencing authority usually rests with judges. Their freedom has been reinforced by statutes giving them wide latitude to set terms and an almost total lack of appellate review of sentence decisions.[6] This discretion has caused more than one judge to criticize American sentencing practices on the grounds that they are, in essence, "lawless."[7]

Professionalism. One reason for the courts' apparent disorganization is that they are staffed by people whose professional norms foster independence of judgment and autonomy. Professionalism presumes a sphere of competence and fosters collegiality. The norms of professionals in general and of the legal profession in particular are, in many respects, antithetical to the norms of bureaucratic organization, which promote hierarchy, supervision, and control, and eschew autonomy.

Superficially, the courts are organized in hierarchical structure, with appellate courts supervising the quality of work in lower courts. But this process of supervision is both passive and expensive, and as a consequence it is not used with great frequency. Those with the greatest stake in appeals, the accused, are typically without resources. If they claim ineffective assistance of counsel—as they very well might—they face overwhelming obstacles to voicing their claims in an appellate court. Similarly, although they are often large organizations, public defenders' and prosecutors' offices rarely closely supervise staff attorneys. Deference to professional judgment is the norm.

While the adversary process itself is often defended as a device for self-regulation, the norms of professionalism remain as the primary constraints on and guides to conduct of courthouse officials. Yet here as elsewhere the weaknesses of professionalism as a system of regulation are widely recognized.

THREE ASPECTS OF FRAGMENTATION

Value conflicts inherent in the criminal law, multiplicity of organizations, and influence of the larger social and political environment—the three aspects of fragmentation—further influence the functioning of the courts.

Value Conflicts. In his book *The Crime of Punishment,* Dr. Karl Menninger argues that imprisonment is barbaric, that it serves no deterrent or rehabilitative value, and that it reinforces criminal tendencies.[8] He articulates a rehabilitative ideal, which substitutes treatment for punishment. In developing this thesis, Menninger expressed views shared by a great many people, including lawyers, judges, and corrections officials.

In contrast, James Q. Wilson in *Thinking About Crime* argues that the rehabilitative ideal is naïve and soft-headed.[9] Instead he offers a deterrent and incapacitative justification for punishment. This position holds that the function of imprisonment is either to serve as an example to deter others from committing crimes or to isolate from the community those who cannot abide by society's rules.

The differences in these positions are profound; in some formulations what each embraces the other categorically rejects. Such value conflicts are important because they inform the views of many who are involved in the day-to-day work of the criminal justice system. And how one views the purpose of the criminal law and the efficacy of sanctions affects decisions made at arrest, charging, sentencing, and parole.

For example, John Hogarth found that judges in the same jurisdiction sentenced similar types of offenders with similar

The Courts and Change

backgrounds to widely varying lengths of imprisonments.[10] His investigation revealed that the reason for this was each judge's own philosophy of punishment. Studies of police arrest practices, prosecutors' charging patterns, and parole board decision making report similar findings.

Still another value conflict contributes to the fragmentation of the courts. In his classic study, *The Limits of Criminal Sanction,* Herbert Packer developed two models of the criminal process, the Crime Control Model and the Due Process Model.

The Crime Control Model emphasizes the need to repress crime by maximizing the effort to locate, apprehend, and convict offenders. This view places a premium on speed, finality, professional judgment, and efficient organization. To this end, it prefers the rapid judgments of police and prosecutors to the slow, cumbersome, and artificial process of decision making in the courtroom.

In contrast, the Due Process Model focuses on "the concept of the primacy of the individual and the complementary concept of the limitation on official power."[11] Due process erects a wall around the individual to protect him against the state's unwarranted intrusion, unreliable judgments, and harsh penalties; and at the same time it places limits on the powers of public officials, even if so doing means that the guilty may escape punishment.

These two models represent degrees in emphasis rather than opposing philosophical positions. Nevertheless, they capture the divergent goals of the various agencies and individuals who constitute the criminal justice system, as well as the goals of those who seek to reform the system.

Multiplicity of Organization. While criminal courts often have been characterized as bureaucratic, in reality they are far from it. Bureaucracy implies rational organization, hierarchical control, common purpose, and central administration. None of these accurately characterizes the courts. Their fragmented organization reinforces the diversity of values and interests within them. No single coordinator, no central authority exists to resolve disagreements or to enforce compliance with goals. No set of incentives binds courthouse officials into a coherent group. Whatever one's objectives in the criminal justice process, there is some leeway and opportunity to pursue them.

Influence of Environment. Martin Levin has found clear differences in severity of sentences between courts in Minneapolis and Pittsburgh. These he explained by differences in the values and perspectives of the judges in those two cities, differences, he went on to show, that stemmed from the political cultures of the two communities.[12] Minneapolis is a middle-class "good government" city; Pittsburgh is a city of ethnic, machine politics. In Minneapolis, few of the judges are drawn from backgrounds in electoral or ward politics; in Pittsburgh, most of the judges received their appointments as rewards for party service. Differences in judicial performance are equally clear-cut. Minneapolis judges are more "legalistic," less likely to plea bargain, and harsher in sentencing; Pittsburgh judges are informal, active in encouraging and taking part in negotiations, and more lenient. Levin argues that these judges act as they do because of their backgrounds and the process by which they are recruited. Pittsburgh judges have lower-middle-class "ethnic" backgrounds and as a result have a greater capacity to empathize with defendants than do the middle-class Minneapolis judges. What Levin found in these two cities, others have found in still other cities. In New Haven, Connecticut, the political culture affects not only judges, but prosecutors and public defenders as well.[13] Studies of courts in Chicago, Baltimore, Detroit, and elsewhere reveal similar patterns.[14]

The Courts and Change

Assessing the Courts

It is precisely the lack of agreement on fundamental goals that leads to vociferous disagreement about progress in the courts. What one person is likely to hail as progressive, another may regard as a step backward. To some, the landmark decisions of the Warren Court on criminal procedure were a step forward; to others they represent capitulation in the fight against crime. The very terms *reform* and *improvement* are subjective.

Whatever one's goals, there is a tendency to expect too much of the courts. Higher standards can lead to improvements, but exaggerated expectations can also foster disillusionment. Given the magnitude of the concern with crime in American society and the limited ability of the courts to deal with this problem, the disjuncture between expectation and reality can lead to a sense of crisis even as things improve, as well as an insistence that the courts solve problems beyond their capacities. Courts cannot solve the problem of crime or even make a significant dent in it. Thus, in a very real sense the courts—charged with handling society's failures—will always fail. What the family, the church, the workplace, and the school cannot do, neither can the courts.

This view requires a realistic appraisal of the capacities of the courts and suggests that courts be understood in historical perspective and social context. Such a view will lead us to abandon many long-standing notions. It will lead us to moderate expectations and sensitize us to the range of changes we might reasonably expect.

This point can be illustrated by examining several recurring allegations about the failings of the criminal courts: (1) plea bargaining signifies a decline in the adversary system; (2) heavy case loads are the major cause of delay and plea bargaining; (3) courts are overly lenient and undermine the deterrent effects of the law; and (4) the low quality of court personnel contributes to the failures of the court.

Plea Bargaining. One of the major charges against plea bargaining is that it is a cooperative practice that, as a result of heavy case loads, has come to replace the combative trial, and as such compromises the integrity and effectiveness of the adversary process. This charge is false. It lacks historical perspective and fails to place plea bargaining in the proper context.

Negotiations have long been a standard feature of criminal process, and reliance on bargaining certainly predates the spurt in the crime rate of the past two decades. Negotiations in the criminal process have their origins in the Middle Ages if not before.[15] In America, plea bargaining was common in colonial New York[16] and well established shortly after the Civil War.[17] In the 1920s, guilty pleas in Cleveland accounted for 86 percent of all convictions; in Chicago, 85 percent; in Des Moines, 79 percent; in Dallas, 70 percent.[18] These high rates of guilty pleas were found not only in the crowded urban courts, but also in less crowded rural courts. In rural New York State in the early 1900s guilty pleas accounted for 91 percent of all convictions, higher even than the 88 percent rate for New York City.[19] Two recent studies of Connecticut's courts found that, while trial rates have varied, they have hovered around 10 percent for the past ninety years.[20] Other historical studies of criminal courts paint similar portraits.[21]

These findings must be interpreted with caution. The absence of a trial does not by itself indicate a plea *bargain,* an exchange of a guilty plea for a guarantee of leniency. Still, the relatively low rates of trial clearly disprove the charge that there was a more effective adversary system in the past. At least as measured by the form of disposition during this century, we are not experiencing a dramatic weakening of the adversary system.

20

The Courts and Change

The trial records of courts in Connecticut, Wisconsin, New York, and London's Old Bailey from the early nineteenth to the mid-twentieth century[22] do not reveal a vastly superior system. Although there were more trials at one time, all but an occasional handful bear scant resemblance to the popular image of a vigorous duel between skilled adversaries. Often, one judge and one jury heard five or six cases or more in a single day, averaging perhaps three-quarters of an hour per case from charge to verdict to sentence. Typically, the prosecutor would summarize the charges and introduce the complaining witness, who would tell his or her story quickly. Usually, defendants who were unrepresented by counsel remained silent or stammered a sentence or two of defense. After a brief deliberation, the jury would give its verdict, and if the verdict was guilty the judge would immediately pronounce sentence. The few rules of evidence and procedure were usually honored in the breach.

Other reports on trials during the nineteenth and early twentieth centuries paint much the same picture. Stanford law professor Lawrence Friedman has sifted through turn-of-the-century trial records of American courts. In Oakland, he found that many cases were handled in a perfunctory manner. In a Leon County, Florida, court, "there were as many as 6 'trials' a day in the 1890s, complete from selection of a jury to verdict. Yet the court handled other business as well."[23] One need not point to the horrors of occasional perfunctory trials; the evidence on the mundane and routine cases is much more telling.

We see that when trials were more prevalent, they usually took less pretrial preparation and court time than the typical guilty plea does today. Indeed, one significant difference is that today those accused of serious offenses are routinely represented by counsel. And if an appearance in the courtroom today is still little more than a brief ritual, at least both prosecutor and defense attorney are likely to have the resources to review the case, undertake some investigation, and debate and negotiate the issues prior to this appearance. Viewed in this light, plea bargaining can be seen as an expansion of adversariness, and certainly not its demise.

If the trials of the past fall something short of the battles they sometimes are envisioned to be, earlier methods of arriving at guilty pleas are even more revealing. Even in the 1950s, observers could report that unrepresented defendants accused of serious crimes were threatened by fast-talking prosecutors to plead guilty or face harsher treatment before a jury:

> The methods used by the prosecutor and the judge to obtain a plea of guilty to a lesser charge from an unrepresented defendant often amount to downright coercion performed in open court. I have heard one prosecutor tell a defendant, "Don't be a fool—if you buck us you will wait six months in jail for your trial. Now if you take a plea, you'll get six months and at the end of that time you will be a free man." Another prosecutor told an unrepresented defendant, "You had better plead guilty to petty larceny or we'll make sure you are sent up for ten years in the penitentiary."[24]

Such reports reveal not the adversarial ideal, but something that is difficult for the contemporary observer to comprehend: *arrestees unrepresented by defense counsel* were typically rushed through crowded and noisy courts and pressured to plead guilty by prosecutors—and those practices were condoned by judges. Even in those cities that had developed public defender systems before the 1963 Supreme Court decision requiring routine appointment of counsel in felony cases,[25] public defenders were available in only special circumstances, and then only after indictment, by which time the vast majority of cases had already been terminated by pleas of guilty. Seen from this perspective, the presence of a defense attorney who is able to review the case and negotiate with the prosecutor constitutes a substantial strengthening of the adversary system.

The claim that plea bargaining signifies a decline in the adversary process is a rhetorical appeal to a mythical yesterday, a state-of-nature fallacy all too common in the analysis of social problems. If plea bargaining is an indicator of the twilight of the adversary system, when was its high noon? These observations are not offered to justify the intimidation inherent in plea bargaining, but to place this practice in context and in so doing question conventional solutions to a serious problem.

Heavy Case Loads. Between 1960 and 1970, the arrest rate jumped from 3.7 to 6.6 million, an increase that was regarded as epidemic. At the same time, there was an increased concern with the growing problems of backlog and delay. The conventional response was that more money was needed, a position that implies that these problems would diminish if enough resources were provided.

However, it has not been demonstrated that heavy case loads necessarily cause backlog and delay. Comparisons of practices between heavy and light case load courts are revealing. In both, guilty pleas predominate, and there are few salient differences in the ways cases are handled. Perhaps most significant is that judges and prosecutors in the heavier case load courts tend to put in longer hours than do their counterparts elsewhere.[26]

There are various ways case loads can affect court practices. In 1972, the number of judicial "parts" (a judge, plus clerks, stenographers, prosecutors, and so forth) in New York State was substantially increased in an effort to cope with the mounting backlogs. One official survey found that the expected reduction in delays had not materialized. Commenting on the survey, the New York Commissioner of Criminal Justice Services concluded:

Although 51 new felony trial parts have been added since September 1972, and the number of defendants indicted has decreased substantially, *the number of defendants waiting trial has declined only slightly.* In addition, the proportion of defendants whose cases are disposed of by guilty pleas has declined, while the proportions of those disposed of by dismissal and trial, respectively, have increased. The data indicate clearly that the expenditure of substantial sums for additional trial parts has not solved the problem of delay and backlog.[27]

Others have reached similar conclusions. An examination of a second round of increases in judgeships in New York State concluded that "backlogs have [increased] in spite of the addition of 31 new judges assigned to deal with new law cases, furnished at an annual cost of $23 million."[28] A report on the criminal courts in Washington, D.C., found that *additional resources* had the effect of *increasing* delays and pressure to plea bargain: "[With the addition of new judges] . . . the average number of misdemeanor cases disposed per judge dropped from 23.5 cases to 18.9 cases."[29]

Many discussions of the case load problem assume that the courts are being swept under a rising tide of work, and in terms of numbers of cases, this appears true. There is more reported crime today than ever before,* but the courts have fared well in obtaining resources proportionate to these increases. While arrest rates increased by 290 percent between 1950 and 1970, during the same period expenditures for the criminal court sys-

---

*Contrary to the alarm of a "law explosion" in recent years, the problem of heavy case loads facing the courts is not new, nor is it the product of the liberal Warren Court, rampant criminality, declining morality, or some other such evil. To the extent that it exists, it is an age-old problem. The Bible records: "Moses sat to judge the people . . . the people stood by Moses from the morning unto the evening," because he could not judge all the cases by himself. In response to complaints by Moses, Jethro, his father-in-law, recommended that he train others in the law and delegate them the small matters, reserving the hard cases and appeals for himself. *Exodus* 18: 13–27. Contemporary commentary on English courts from the Middle Ages to the present is filled with complaints about the overworked and careless judges.

The Courts and Change

tem increased by 325 percent.[30] Expenditures for the judiciary increased at an even higher rate, 500 percent, between 1954 and 1975.* Furthermore, while there are indications that the increase in serious crime has slowed down since 1975, expenditures have continued to climb; yet problems of delay have increased. This suggests that the case load problem involves much more than sheer numbers and dollars. According to the diagnosis of former Attorney General Ramsey Clark:

> At the federal level from 1956–1958 there were never fewer than 30,000 or more than 34,000 criminal cases initiated in any year. The nature and complexity of the cases changed, but their number was fairly stable. The number of judges handling the cases, on the other hand, increased 45 percent, while the number of Assistant United States Attorneys increased only 14 percent. Simultaneously, the number of criminal cases pending—the backlog—increased by more than 100 percent. With twice as many cases pending, only a few more new cases commencing, and far more judges to handle them, it became clear that the constitutional right to speedy trials would depend on more than just additional judges. *The solution must include better techniques and*

---

*Criminal justice expenditures for all levels of government, federal state, and local, increased from more than $2 billion in 1954 to $3.3 billion in 1960. In 1965, total expenditures reached $4.6 billion and in 1970, $8.6 billion. As of 1975, total expenditures were over $17 billion. Looking solely at judicial expenditures, one sees a similar progression. In 1954, slightly less than $400 million was spent on the judiciary. By 1960, the figure was $597 million; by 1965, $748 million. In 1970, spending on the judiciary was just under $1.2 billion, and by 1975, had gone over $2 billion. Arrest data show comparable growth rates. From 1954 to 1960, arrests increased two and one-quarter times, from 1.7 to 3.7 million. From 1965 to 1970, these levels expanded by 30 percent, from 5 to 6.6 million, and in 1975, arrests went over 8 million for the first time. Thus, between 1954 and 1975, expenditures for the judiciary increased over fivefold, while arrest rates increased at just under fivefold.

For additional arrest and expenditure data, see U.S., Department of Commerce, *Historical Statistics of the United States: Colonial Times to 1970* (Washington, D.C.: U.S. Government Printing Office, 1976), Part 1, pp. 415–16; U.S., Department of Justice, *Sourcebook of Criminal Justice Statistics—1977* (Washington, D.C.: U.S. Government Printing Office, 1978), p. 44; U.S., Federal Bureau of Investigation, *Uniform Crime Reports—for the United States* (Washington, D.C.: U.S. Government Printing Office, 1975).

more supporting personnel, in all activities of the judiciary, civil and criminal. [31] [Emphasis added.]

The solution Clark refers to is essentially the same one offered in the reports on the New York and Washington courts discussed above: "Even if large additional sums could be found during the current difficult period, prudence would still dictate a cooperative effort to find better ways to organize felony case processing."[32] Others have argued that the solution lies in altering "local legal culture" and traditions of the local defense bar rather than adopting any specific administrative device or rule.[33]

No one denies that delay is a serious problem. The preceding discussion suggests that this problem often has not been taken seriously enough. Additional resources alone are not likely to make any difference. More effective would be improved management techniques and programs that alter incentives, informal norms, and long-standing traditions of local defense bars.

Leniency. Complaints about judicial leniency periodically lead to crusades to force judges to dispense harsher sentences. Leniency is, of course, a relative term: one person's leniency is another's harshness. Moreover, sentence reform historically has been cyclical in nature, embracing philosophies of fixed, and then discretionary, sentencing.

The leniency debate may be based upon differences in philosophy of punishment, or it may be based, as is so often the case, upon the extreme example that is reported in the media. It may also be motivated by a concern for enhancing the deterrent capacity of the criminal sanction. Still, it is possible to frame questions about leniency in a meaningful way, by contrasting sentences in the United States to those in other countries, exploring actual reasons for downgrading charges in

The Courts and Change

American courts, and then by examining evidence on sanction severity and deterrence.

In contrast to courts in Europe, sentences handed down in this country are draconian. Sentences in the Scandinavian coun-tries, Holland, Great Britain, and Germany are typically much shorter than those for comparable offenses in the United States.

This comparison will probably not comfort most critics of American courts who focus on the "deterioration" of cases as they wend their way through the system. The process is often described as the "sieve" or funnel effect because of the dra-matic falloff at each stage. For example, one recent study of the criminal courts in New York City found that:

- 80 percent of all *reported* felonies did *not* lead to arrest.
- 43 percent of the felony cases that reached court were *dis-missed.*
- 2.6 percent of all cases were disposed of by trial.
- 74 percent of the felony charge guilty pleas were to misdemea-nors or *lesser offenses.*
- 50 percent of the guilty pleas led to probationary sentences in-volving *no jail time.*
- 41 percent of the guilty pleas led to prison sentences of *less than one year.*
- 9 percent of the guilty pleas led to sentences *over one year.* [34]

Taken at face value, such figures suggest either that the police are careless and arrest people without rhyme or reason or that the courts are incompetent or overworked. There are a number of reasons why cases are filtered out of the court system by dis-missals and pleas to reduced charges, and upon inspection many of them are compelling.

The study by the Vera Institute of Justice presented some startling and revealing facts about the deterioration of criminal cases. Was slippage, it asked, due to carelessness, uncaring prosecutors, crowded calendars, or soft judges? Or was it due to poor police work? The Vera researchers tracked a large sample of cases through the criminal justice system from arrest to sentencing. They found no evidence of widespread failures by the police, nor did they find prosecutors and judges who had caved in under the pressures of congested courts. They discovered, in fact, a group of men and women whose work yielded up roughly consistent and evenhanded decisions. Seemingly arbitrary dismissals, delays, charge reductions, and variations in length of sentences were usually explained in terms of legally relevant factors. In particular:

> The study found an obvious but often overlooked reality: criminal conduct is often the explosive spillover from ruptured personal relations among neighbors, friends and former spouses. [Felony] cases in which the victim and defendant were known to each other constituted 83%, of rape arrests, 69% of assault arrests, 36% of robbery arrests, and 39% of burglary arrests. The reluctance of the complainants in these cases to pursue prosecution (often because they were reconciled with the defendants or in some cases because they feared the defendants) accounted for a larger proportion of the high rate of dismissals than any other factor. . . . Judges, prosecutors, and in some instances police officers were outspoken in their reluctance to prosecute as full-scale felonies some cases that erupted from quarrels between friends or lovers. . . . Thus where prior relationship cases survived dismissal, they generally received lighter dispositions than stronger cases.[35]

The other factor that influenced convictions and sentences was prior record. For example, 84 percent of those who had prior records and were convicted received prison sentences, as opposed to 22 percent of those without prior records.[36]

Prior relationship and prior record are the two principal determinants of how a particular felony case is handled, whether

The Courts and Change

the case survives and, if it does, whether it results in a prison sentence. The Vera Institute's research team found it useful to distinguish between "technical" and "real" felonies, based in large part upon the presence or absence of a prior relationship between the complainant and the suspect.

A second major complaint about lenience is that the courts fail to impose severe enough sentences to deter crime. Typical of such criticism is the position of former New York City police commissioner Patrick Murphy, who has repeatedly asserted that, because of lenience, the courts must assume "the giant share of the blame [for the] disturbing rise in crime."[37] This diagnosis leads to an obvious and popular prescription—increased sentences.

Concerned with just this issue, the National Academy of Sciences convened a panel of distinguished statisticians, econometricians, and criminal justice scholars to review the available evidence on the relationship between sentence severity and the incidence of crime. The panel's report concluded that it could not definitely establish that there is any direct relationship between severity of sanction and deterrence.[38]

This conclusion seems to fly in the face of common sense until one considers that increasing sentences does not affect *likelihood* of apprehension and punishment. Of all crimes committed, only a portion are even reported to the police; of those reported, only a small fraction lead to arrests; of those arrested, only a fraction are convicted. Unless rates of reporting, apprehension, and conviction are all significantly increased, a measurable impact of the threat of an increased criminal sanction at this last stage is remote.

Problems with the deterrent effects of sanctions are compounded in still other ways. One view of punishment holds that, at a minimum, incarceration incapacitates at least one criminal offender and thereby reduces crime rates to the extent that this person is not committing a crime he or she would commit if free. The Academy's report questions even this seemingly obvious fact, pointing out that for some types of criminal activity incapacitation of one person may simply lead to replacement by another. For example, if there is a strong demand for drugs and a large labor market of users to tap, incarceration may have virtually no net effect. Similarly, if crimes are the results of group efforts, as much juvenile crime is, the removal of one member of a group may not affect the criminal activity of the group as a whole.[39]

The lessons are clear, if disillusioning. If we cannot assert with confidence that changes in something so significant as sentence severity will have measurable deterrent effects, we must be even more cautious in our expectations about the effects of other more modest changes in the courts. Still a great many reformers justify all types of minor changes on just such grounds, a practice that if taken seriously—as it has been —leads to unrealistic expectations and disappointment.

Personnel. In assessing the functioning of the courts, we must ask, finally: Are they staffed with unqualified, ineffective personnel?

If one takes a historical view, the answer is clearly no. The practices and the quality of personnel today are likely to be found more acceptable than ever before. A few short years ago, open appeals to friendship, party loyalty, and bribery were not at all uncommon as ways of disposing of cases in the trial courts.[40] Now such practices are limited to isolated cases in a few cities or restricted to traffic tickets. In addition, differences between publicly appointed and privately retained defense counsel are more rhetorical than real. A study comparing case outcomes and sentences of public and private defense attorneys in three cities found:

The Courts and Change

> There was very little difference in the case outcomes received by clients of different types of lawyers, . . . [and that there] is a pervasive antipathy of unexpected magnitude toward publicly paid defense lawyers, primarily those who work for the first-line defender system, and it rapidly became evident that these antipathies were based on defendants' suspicions about the loyalties and the abilities of these lawyers.[41]

While problems with public defenders remain, few other public service institutions can match their records. Few public institutions can boast that, on average, they provide services as good as those that are available in the free market.

Conclusion

Three conclusions emerge from this examination of the criminal courts. First, the fragmentation and seeming inefficiencies of the courts are inherent in the very theory and structure of the adversary process and are not simply the result of aberration, overload, or inadequate personnel. Second, many concerns that give rise to feelings of decline and decay and a sense of crisis in the courts rest upon heightened standards and unrealistic expectations, not actual deterioration and demise. Third, a realistic assessment of the functioning of the courts reveals that, despite continuing serious problems, defendants are better served by the courts today than ever before. However slow and arbitrary criminal courts are today, they are more even-handed and deliberative than they once were.

Part of our disillusionment is due to overexpectation. Courts are faced with a serious crime problem, but they cannot reasonably be expected to reduce crime. With but few exceptions society has not been prepared to have courts impose penalties draconian enough to make a significant difference, and courts have little to do with affecting likelihood of apprehension, the driving force in deterrence. Perhaps if we agreed to lock up selected groups of repeat offenders for very long periods of time, this incapacitation would reduce crime. But recently we have not, and once when the English courts did have capital punishment available in all felonies, they rarely used it.[42] Even if the controversial exclusionary rule were reversed to permit the introduction of illegally obtained evidence in court, the law's deterrent effect probably would not be measurably enhanced.[43]

Similarly, courts handle large numbers of repeat offenders, but they cannot rehabilitate. There are few agreed-upon rehabilitative techniques, and at any rate the courts do not administer them.

More realistic expectations suggest a more hopeful picture. Courts are faced with many more cases today than ten, twenty, and fifty years ago, but they are staffed by more judges, prosecutors, defense attorneys, clerks, and bailiffs than ever before. There remain problems with securing defendants' rights, responding to the needs of witnesses and victims, and dispensing evenhanded sentences, but these problems have diminished over time. There has been a demise of trials, but trials were never close to being what they are portrayed as in myth. There are serious inequities in plea bargaining and major problems with delay, but these issues must be understood in the context of expanded rights for the accused, resources for the prosecutor, and the overall professionalization of the courts.

The courts do face real problems—and problems that have not been taken seriously enough. The question is: Can proponents of planned change adequately identify these problems, diagnose them accurately, and make improvements?

Part 2

Some Innovative Programs

Introduction

THE PROCESS OF

PLANNED CHANGE

TRADITIONALLY, studies of planned change have dealt with how policies are made in the legislative and bureaucratic arenas, not with what happens once those policies are adopted and become routine. In presenting the histories of bail reform, pretrial diversion, sentence reform, and speedy trial rules, Part 2 goes beyond the traditional examinations. It focuses on each important stage in the change process: diagnosis, initiation, implementation, routinization, and then the evaluation of each reform effort. Studying the four reforms in this manner, we can assess the success of each more realistically than has so far been done. At the same time, we can learn how the process of change operates in the criminal courts and why it often leads to mixed and confusing results.

Stages of Innovation

Because each stage in the change process has its own distinctive pitfalls, each must be considered separately:[1]

Diagnosis or Conception. Diagnosis is the process of identifying problems and considering solutions. While a business firm has rough indicators of performance, the criminal justice system enjoys no such luxury. Whoever offers a diagnosis is likely to be greeted with suspicion. Different perspectives lead people to identify different problems and suggest different remedies.

Initiation. During initiation new functions are added or practices are significantly altered. This stage requires several decisions: (1) Which of several alternatives will be adopted? (2) How will the program be financed? (3) Where will the program be located?

Even at this early stage serious problems emerge. Many changes in the criminal courts are initiated by outsiders, such as appellate court judges, legislators, and agency heads. But eventually institutions close to the court must assume responsibility for these initiatives, and when they do, the original intent can be neglected or deflected. Avoidance, evasion, and delay are familiar responses to innovation.

Similarly, initiators are not usually financiers, and new programs have a tendency to adapt to the interests and views of those who ultimately pay for them. This fact alone causes serious adjustments away from the original focus of some programs.

Implementation. Implementation involves staffing, clarifying goals, and adapting to a new environment. Ultimately, it is the task of translating abstract goals into practical policies. If the new program or policy is significant, it will disrupt old routines, challenge established authority, and introduce unpredictability into a process that depends on certainty and stability. Tensions will inevitably develop between initiators and implementors; a strategy that maximizes the likelihood of successful initiation—bold language, simplification, and expansive promises—is likely a strategy that undercuts implementation. Change requires coordination and cooperation, but the complexity of joint action can be frustrating. Thus, as the number of participants in the process of implementation increases, the probability of success decreases. In a system as fragmented as the criminal courts, this is perhaps the single largest obstacle to change.

Routinization. Unless a program is intended to be temporary or a single-shot effort, sooner or later it must be routinized. This involves commitment by an institution to supply funding and a physical base of operations. Ultimately, the success of an innovation must be judged by how it performs under this routine rather than under its initial conditions.

It is at this stage that many good ideas go astray. What worked when supported by ample federal funds often fails when supported by fewer local tax dollars. What worked under the heightened conditions of an "exciting new experiment" need not work after the halo has worn off. What worked for a charismatic leader and zealous followers may bog down with an ordinary staff.

Evaluation. Despite all this, new programs are usually assessed during their experimental (the first three) stages rather than their routine periods (the fourth stage). While such evaluation can tell us something about whether an idea can or cannot work, it tells us next to nothing about whether it *will* work. Little is known about the eventual, routine performance of new programs introduced into the criminal justice system.

The Context of Change

Who innovates? Why? What happens when someone tries to put a good idea into practice? What internal and external forces encourage change in the courts? Courts possess characteristics that simultaneously promote and retard change. On the other hand, planned change is most likely to succeed in institutions where

- highly trained professionals perform complex tasks
- authority is diffused and flexible rather than centralized
- duties are left ambiguous rather than formally codified in detail
- roles and mobility are flexible rather than rigidly stratified.[2]

However, two factors have been found to discourage innovation:

- the higher the volume of production, the greater the need for established routine and the lower the incentive to change
- the greater the emphasis on efficiency, the more likely that program change will be discouraged.

The courts are staffed with trained professionals—lawyers—whose loyalty to organization and adherence to bureaucratic routine is tempered by a sense of professional autonomy, a characteristic that gives rise to innovation. In addition, courts are not bound by rigid centralized authority, still another condition that facilitates initiative and fosters innovation.

But courts also are enmeshed in a web of rules that can be and often are inimical to change. Those comfortable with current practices selectively invoke these rules to impede change.

Because courts are rigidly segmented, broad perspectives and systemwide thinking are discouraged and innovation is stifled.[3] Segmentation in the adversary process inhibits communication, feeds distrust, and breeds antagonism.

Finally, because they handle large numbers of cases, courts are forced to emphasize efficiency. This leads officials to stereotype cases and adopt informal practices designed to process

The Process of Planned Change

cases rapidly. In such a system, any deviation from standard practice is resisted because it threatens to disrupt the natural flow of business.

All this is unwittingly reinforced by the adversary theory, which encourages mutual skepticism and divides power and responsibility among participants in the process. When change is initiated, it may not be implemented. One group may authorize a new program, but another may not fund it. Legislators may place new laws on the books, but prosecutors and judges may ignore them. Changes may be internal, affecting only one agency, and not yield any important systemwide effect. Because attempts to change, to innovate, inevitably encounter at least some of these problems, the successes are few and far between.

Chapter 2

BAIL REFORM

Background of Modern Bail Reform

The clash of conflicting interests in the criminal process is nowhere more evident than in the administration of bail. Although the accused is entitled to the presumption of innocence, the public has a legitimate concern in securing his appearance in court and protecting itself from danger. Under early common law, the practice devised to balance these interests was bail, which permitted the accused to offer some form of security as a guarantee of court appearance. The Eighth Amendment elevated this practice to a constitutional guarantee, providing that "excessive bail shall not be required."

The origins of bail are not certain. One view holds that it evolved from hostageship, whereby a willing hostage was substituted for the accused and liable to be tried and punished in the event of the accused's failure to appear for trial. Another theory states that it derives from *wergeld*, the medieval practice that required those accused of defaulting on loans to demonstrate their ability to compensate the lender in the event of conviction.

Whatever the precise origins, bail emerged as a practical ne-

40

cessity. In thirteenth-century England, a formative period in the development of bail, judges rode circuit, and it could be many months before cases were tried. Pollock and Maitland observe in their classic history of English law that despite the delay, it was not common to imprison the accused. This was "not due to any love of an abstract liberty," but because "imprisonment was costly and troublesome."[1] Until the mid-nineteenth century in England and the United States, those accused of noncapital offenses were routinely released into the custody of someone who would vouch for them or after pledging their own property as surety.

Changing conditions in the nineteenth century disrupted this tradition. The rise of a propertyless class and increased mobility complicated suretyship. The accused often did not have property, and, increasingly, flight abroad or to the frontier became practical alternatives to standing trial. The result was increased use of the jail, a facility created to detain those accused of crimes. Thus, as flight became a practical alternative, the state relied more and more on pretrial detention.

This development fostered the rise of the bail bondsman, a businessman who, for a fee, assumes financial responsibility for the accused and guarantees his appearance in court. By the end of the first quarter of the twentieth century, courts in major American cities routinely set bail beyond the reach of the accused and in doing so provided windfall profits for a few bondsmen and often substantial side payments to cooperative jailers and judges.[2]

Thus, ironically, as the formal right to bail crystallized, conditions for securing pretrial release became more stringent and releases actually declined. Pretrial release was routine through much of the nineteenth and early twentieth centuries, but by 1950 almost everyone arrested on a misdemeanor or felony charge had money bond set. Two surveys taken in the early

1960s found that more than three-quarters of all felony arrests required bail of $1,000 or more, while two-thirds of all misdemeanor cases involved bond of $500 or more.[3] Given the standard bondsman's fee of 10 percent, this meant that pretrial freedom cost $50, $100, or more in cash on the barrelhead, as well as the posting of collateral.[4]

Discovery and Rediscovery of the Problem of Pretrial Release

Like so many social problems, bail and pretrial detention have a long history of being discovered only to be forgotten. In the mid-nineteenth century, Charles Dickens wrote about usurious bail bondsmen in England, and contemporary descriptions of American bondsmen were similar.

Early in this century, Roscoe Pound and Felix Frankfurter studied the administration of the criminal process in Cleveland and other large cities and found that the "real evil in the situation is not the matter of easy bail, but the professional bondsmen who make a business of exploiting the misfortunes of the poor and whose connections with 'runners and shysters' tend to prostitute the administration of justice."[5] The first major empirical study of bail administration was conducted by Arthur Beeley of the University of Chicago,[6] who uncovered the same types of questionable and corrupt practices found in Cleveland. His 1927 book, *The Bail System in Chicago,* reported that arrestees were held for long periods before formal charges were filed, were not allowed to contact friends or relatives, and were refused release on cash bail or recognizance. The amount of bail was determined by the seriousness of the charges rather than by the defendant's ability to pay or by his ties to the community. Finally, more than half of those accused of a crime were never convicted, and fully one-third of those jailed until disposition were not convicted. Other studies followed, revealing similar patterns in other communities.

The prestigious Wickersham Commission issued its final report in 1931 and drew heavily on Beeley's and others' bail studies.[7] In response, Attorney General Cummings proposed that Congress attempt to improve the administration of state and local criminal justice. But Congress never mounted a national effort.

One consequence of these early studies was the regulation of bondsmen, but, as with so much regulation in general, those to be regulated themselves shaped the legislation. These regulations made entry into the business difficult, protected entrenched bondsmen from competition, and provided ways for bondsmen to avoid full liability for clients who absconded. If anything, these provisions enhanced the significance of money bail and bondsmen in the criminal process.

The depression and war contributed to the decline of interest in bail reform, and it was not until concern over class and racial inequity emerged in the 1950s that bail resurfaced as a social problem. In a series of landmark cases, the United States Supreme Court, under Chief Justice Earl Warren, first tackled the problem of racial inequality and then expanded the rights of the criminally accused.[8] It was nearly inevitable that these concerns would lead to pressure to reexamine the constitutional issues of bail as well.

One of the first to initiate bail reexamination was University of Pennsylvania law professor Caleb Foote, who studied bail administration in Philadelphia and New York City. Little had changed since Beeley, and, for all practical purposes, Foote's reports could have been written twenty-five or fifty years earlier. However, his studies carefully documented what had only been hinted at before, that those who cannot afford bond are much more likely to be convicted and to serve longer sentences than those who can.[9]

Foote's work set the agenda for another bail reform effort, which he anticipated would take place in the courts as a response to litigation seeking clarification of the Eighth Amendment's bail provision. His program was essentially the same one that had been proposed thirty years earlier. It sought to reduce:

- class bias of bail, which allows the well-off to purchase their freedom while denying opportunity to the poor
- harsh treatment and penalties for those unable to secure pretrial release
- delays in securing pretrial release
- costs of obtaining freedom
- corruption in bail administration (whereby bail bondsmen engage in kickbacks to jailers and avoid paying forfeitures in instances of absconding)
- high rate of failures to appear in court as a result of inept court management

Recent Efforts: The Vera Institute and Aftermath

This time the agenda remained, and substantial changes have occurred, although Foote's anticipated "constitutional crisis" in bail administration never did materialize. Instead, the reforms that he helped launch are primarily the result of an administrative revolution. Altogether, judges have remained strangely silent on bail reform.

Two of the earliest advocates of this administrative approach to bail reform were nonlawyers: Herbert Sturz, a young social worker and magazine editor, and Louis Schweitzer, an elderly industrialist and philanthropist, who began working together in 1960. Shocked by the conditions he found on a tour ar-

Bail Reform

ranged by Sturz of New York City's pretrial detention facilities, Schweitzer resolved to establish a nonprofit bail loan fund. Instead, Sturz suggested that Schweitzer provide funds to establish a pretrial release agency, which would help arrestees secure release on their own recognizance, altogether circumventing the need for money bail. Schweitzer agreed and pledged $50,000 to support the project. This was the beginning of the Vera Foundation (later the Vera Institute), which has since become synonymous with criminal justice reform in general.

With support from the politically connected Schweitzer, Sturz convinced city officials to allow him to set up a short-term experiment. He would check the backgrounds of arrestees as soon after arrest as practical, and if they revealed close community, family, and employment ties in the city, he would recommend that they be released on their own recognizance (ROR). This would aid the accused by securing their early release without the need for bond and aid the state by reducing costly detentions.

The experiment began operations in Manhattan in the fall of 1961. Staffed by student volunteers from New York University School of Law, it interviewed arrestees, verified information on their backgrounds, and then, at arraignment, if the information warranted, recommended ROR. Vera's volunteers remained in contact with releasees, telephoning to remind them of subsequent court appearances.

After its first year, the Manhattan Bail Project, as the effort came to be known, sought to evaluate its own effectiveness. It compared those who had been recommended for release and had actually been released with a control group, which consisted of arrestees who had not been recommended or released. Vera's study purported to reveal overwhelming and persuasive evidence of the viability of the project.

From the outset, officials were fearful that the project would precipitate a flood of no-shows in the court. The study spoke to this issue as well; failure-to-appear (FTA) rates among project participants were, on average, lower than those of the control group who had posted bond.[10] The project reported that of the first 250 RORs, only 3 failed to appear in court, a success rate dramatically higher than with other types of release.

These claims served both to stimulate nationwide interest in pretrial release and to set the course for bail reform. Rather than challenging the constitutionality of a money bail system that discriminated against the poor, the movement adopted an administrative orientation: it created an institution to administer ROR programs.

During the several years following this initial effort, interest in bail reform grew at an exponential rate. Additional support from the Ford Foundation allowed Vera to expand its pretrial release operations to other boroughs and to aid civic leaders in Des Moines, San Francisco, and other cities in establishing ROR projects.

Even before the Manhattan project had gotten under way, Sturz was asked by Dean Francis A. Allen of the University of Michigan Law School and head of a federal commission on criminal justice to undertake a comprehensive examination of bail in the New York City federal courts. Sturz issued a highly critical report, which aroused the interest of Attorney General Robert F. Kennedy. This contact with Kennedy eventually led to the 1964 National Conference on Bail and Criminal Justice, which launched the bail reform movement in earnest. Jointly sponsored by the Department of Justice and the Vera Foundation, and organized by Justice Department officials Daniel J. Freed and Patricia Wald, the conference brought together more than four hundred judges, prosecutors, defense lawyers, police officials, bondsmen, and other interested persons. Freed and Wald wrote a monograph, *Bail in the United States,* in connection with the conference. Eloquent, brief, and widely disseminated, this document provided a plan for bail reform that came to be accepted across the country.

Shortly after the conference, Kennedy directed all U.S. attorneys to release on recognizance as many of those accused in the federal courts as appeared practical. In 1966, Congress enacted the comprehensive Bail Reform Act, the first major piece of legislation affecting pretrial release since passage of the Judiciary Act of 1789. Extremely liberal in encouraging nonmonetary release, the act served as a model for many state reform acts.

Replicating the Reforms

The enthusiastic response to the Vera Project, the replications in Des Moines and elsewhere, and the publicity accorded to the conference all helped precipitate widespread interest in bail reform; within eighteen months after the 1964 conference, 61 new pretrial release projects had sprung up around the country. By 1969 this number had jumped to 89, and in 1973 the number of projects was 112.[11] It is estimated that there are now well over 200 specialized pretrial release projects. Vera's original Manhattan project served as the model for most of these new programs; this means that many are independent agencies the courts contract for services.

There have also been other important efforts. Some jurisdictions have attempted to modify rather than eliminate money bail by instituting a "10 percent bond program," which allows the accused to post a returnable deposit directly with the court. This reform is generally regarded as successful. There have been a few efforts to develop supervised release programs, whereby an arrestee without roots in the community can be released into the custody of a third party who assumes responsibility for his court appearance. What little evidence there is suggests that this effort has had only mixed success.[12]

The most frequently used alternative to bail is the use of police summonses and citations. A law enforcement officer issues a court summons or citation—a ticket—at the site of the arrest, thereby avoiding booking, arraignment, and bond altogether. Long used in various code violations and minor motor vehicle cases, in recent years summonses and citations have been used as well in petty criminal offenses.[13]

Assessing a Decade of Effort

Unlike the interest in bail in the 1920s and 1930s, the concern that emerged in the 1960s and 1970s was translated into action that has yielded tangible results. Today pretrial release policies and practices are markedly more liberal than those of just a few years ago, and these changes are directly attributable to the efforts of Caleb Foote, Herbert Sturz, Louis Schweitzer, Robert F. Kennedy, Daniel J. Freed, and Patricia Wald.[14]

Wayne H. Thomas examined changes in pretrial release practices in a number of large cities during the decade 1962–71 and found that ROR had become a common form of release.[15] Furthermore, he found that on average, the dollar amount of bail had declined. In short, there had been a dramatic increase in the proportion of arrestees able to secure release prior to trial. Progress, however, has not been smooth. Liberalizing pretrial release increases the problems of nonappearance and rearrest on other charges. Still, only a few bail violators become real fugitives from justice; most eventually return.

It was probably inevitable that opponents of liberal pretrial release would oppose the release of allegedly dangerous people, so often those for whom high bail is set. In late 1969, after a five-year period of liberalization of pretrial release, the new Nixon administration proposed a series of revisions of the federal criminal code, including one that provided for preventive detention. This was adopted only for the District of Columbia but until quite recently has lain nearly dormant.[16] In the early 1980s states began adopting similar provisions, although as of this writing, there has been little experience to determine how frequently they will be invoked.

Causes of Change: The Vera Experiment

Herbert Sturz quickly shaped the Vera Institute into a vibrant and innovative pretrial service program for New York City's courts. The Vera staff developed the standard operating procedure that was subsequently followed in a great many other pretrial release agencies: the project staff inquired whether incoming arrestees wanted to be interviewed for ROR consideration. Those who consented were then interviewed for about ten minutes about their prior records, family ties, employment or school status, and length of residency in the city. Their responses were coded according to a standard form. For instance, someone with no prior convictions was given one point, while someone with three or more misdemeanor convictions lost two points. A total of five points was required for Vera to recommend ROR, although Vera took no action until the information was verified. The staff remained in contact with all of the arrestees who were subsequently recommended for and received ROR, periodically telephoning reminders of future court dates and tracking down those who failed to respond. The distinctive feature of this procedure, Vera and its many supporters claimed, was the rational process of developing and applying predictors of appearance based upon the strength of the defendant's ties to the community.

Vera conducted an experiment, comparing its clients with a control group. A series of reports suggested that Vera's project was an overwhelming success: 59 percent of the ROR recommendations made by Vera were followed by the court, while only 16 percent of those eligible in the control group, but not recommended, were released; almost 60 percent of those whom Vera had recommended for ROR and who were actually released had all charges against them dropped, in contrast to only 23 percent of the control group; and only 21 percent of the recommended and released defendants who were subsequently convicted went to prison, in sharp contrast to 96 percent of those in the control group. Finally, the study reported that in its first three years, only 1.6 percent of those released on its recommendations failed to appear in court, in contrast to an average nonappearance rate of 4 percent for those released on bail.[17]

These and similar figures were presented as dramatic evidence of Vera's success and were quickly picked up by proponents of pretrial release elsewhere.

There are, however, a number of serious problems with the design of this study. The data on the two groups were not perfectly comparable: at times the "best" of the experimental (Vera) group was compared with the "worst" of the control group. Given this, it is not surprising that the project appeared to fare so much better.[18] The more valuable test of the project would have been to compare the release rates, failure-to-appear rates, case outcomes, and sentences of all those initially included in both groups. While the magnitude of the differences suggests that in all likelihood the project had significant positive effects, the research methods were faulty.

A central claim of the research reports was that the "community ties point scale" was a good predictor of future court appearance, and the project's low FTA rate was offered as evidence to support this claim. The scale allowed for substantial discretionary judgment, so that in fact it is impossible to know just how valid and reliable the "objective" community ties factors were. It was complicated by still another factor: whenever an arrestee was released through the auspices of Vera, the project remained in contact with him by mail, telephone calls, and personal visits. It may very well be that the higher appearance rate for project releasees was due not to the community ties predictors, but to this follow-up effort.

Because they were pioneering, the research reports of the Vera Foundation were widely read and well received and convinced a host of funding agencies to expand the project in New York and initiate similar efforts elsewhere. Years later these same flawed studies continue to be cited by proponents of pretrial release programs.

Whatever the shortcomings of this research, Vera's pretrial release project continued to prosper. Between 1961 and 1963, it extended its hours of operation and expanded into the other boroughs; its staff grew, and its budget—supplemented by additional funds from the Ford Foundation and contracts with the city—mushroomed. But despite its energetic director, an enthusiastic staff, and the support of highly placed public officials, Vera's pretrial release project continued to operate in ad hoc fashion, dependent upon informal understandings with the courts, and without any permanent base of funding.

Institutionalizing Pretrial Release in New York City

The impetus to institutionalize came in 1963 as a result of disturbances in the Tombs, the city's pretrial detention facility.

Mayor Robert Wagner announced that the city would take over and expand Vera's pretrial release project as part of a plan to relieve jail overcrowding. In 1964, the project was transferred to the city's Office of Probation.

Vera was an independent, nonprofit agency that had supported its operations through private funds. Its sole purpose was to promote ROR in a skeptical, if not hostile, world. The Office of Probation was part of that world and, in the view of many, part of the problem. Staffed by career officials in an enormous city bureaucracy, this agency had many different responsibilities and no special commitment to increasing pretrial release. It is not surprising that transfer led to a number of important changes.

When Vera was not able to recommend an arrestee for ROR, it had remained silent. In contrast, Probation gave a negative recommendation. The office also altered Vera's point scales, making it more difficult to qualify for an ROR recommendation. Vera had committed substantial resources to its follow-up and notification program. In contrast, the Office of Probation focused almost exclusively on increasing the number of initial interviews and made little effort either to verify information or to notify releasees of future court appearances.

This appears to have been deliberate. A 1967 study by Andrew Schaffer, an attorney on the staff of the Vera Institute, found that over 75 percent of the interviews were not verified, and that only 43 percent of the verified cases led to recommendations for ROR. Thus, out of every 100 interviews conducted by the office, only 25 led to positive recommendations for ROR, and fewer than half actually obtained ROR.[19]

As the Probation-run release project expanded, it accomplished less and less. Despite a tenfold increase in the number of interviews (from 11,556 in 1964 to 136,890 in 1970) and a fourfold increase in budget, ROR releases attributable to the

Office of Probation decreased, as a proportion both of those interviewed and those released.

Soon only a very small portion of all arrestees released on their own recognizance (less than 17 percent) actually gained their freedom with the help of the Office of Probation. Indeed, by 1967, judges were accepting only 32 percent of the Office's positive recommendations, while at the same time releasing fully 28 percent of the office's negative recommendations.[20] This low rate contrasts sharply with the earlier 70 percent acceptance rate of Vera's recommendations.[21]

Whatever the merits of the Probation-run ROR effort, it appears that its expansion and budgetary increase were due to something other than its ability to maintain, let alone improve upon, the accomplishments of the pilot effort. Numbers of interviews rather than release rates or failure-to-appear rates impressed city budget officials.

Re-examining the City's Commitment to ROR:
A Quasi-Experiment

Neither Herbert Sturz nor the Vera Institute lost interest in ROR after the transfer to the Office of Probation. Vera commissioned a study of the operations of the Probation-run ROR program. When the report revealed serious shortcomings, Sturz proposed that the project be transferred back to Vera. Faced with the need to trim its budget in light of the city's fiscal crisis of 1973, the Office of Probation did not resist and the program was quietly transferred, this time supported by funds from the Law Enforcement Assistance Administration (LEAA).

This second round of the Vera-run pretrial release enterprise began operations in Brooklyn in June 1973, under the name Pretrial Services Agency (PTSA, pronounced "Peet-sa"). It expanded to Staten Island and the Bronx in December and to Manhattan during the next two years. As in the earlier Vera-administered effort, the basis for pretrial release recommendations was to be the accused's ties to the community.[22]

The agency redesigned the point scale and developed a research capacity "to assist the judges in evaluating the community ties through the use of statistical analysis."[23]

PTSA quickly rekindled the energy and experimental attitude that had characterized Vera's initial effort. It was aggressive in obtaining federal funds. It revitalized the moribund notification system. It developed a sophisticated computerized information system. It experimented with a program of supervised release. And it retained the noted sociologist Paul Lazarsfeld to assess its community ties point scale.

Lazarsfeld found that the vast majority of arrestees had substantial ties to the community and recommended that PTSA engage in more sophisticated research and design a simpler community ties index. Most of Lazarsfeld's recommendations were put into operation, although the New York Legal Aid Society and the city's judges would not allow PTSA to conduct experiments using random assignments. Like Vera's earlier efforts, it appeared that the project made a difference, although the study was unable to clarify which facets of the project accounted for the difference.

Ten years after the first pilot pretrial release program, and after dozens of studies, the same questions remain unanswered: Is the community ties point scale a valid and reliable predictor of future appearance? Are lowered failure-to-appear rates due to this predictor or to subsequent follow-up efforts? To what extent do the "successes" of pretrial release programs stem from their ability to anticipate what judges will do rather than

Bail Reform

from their "independent" predictions? Finally, there is the nagging but ignored problem of violent arrestees. The vast majority of those released on low bail or ROR are charged with petty offenses that involve no violence. The question facing PTSA and the courts is what to do with someone charged with a violent crime who scores high on the community ties index. To date, PTSA, like other pretrial release agencies, has been content to recommend release but not fight when its recommendations have been disregarded.* This stance is understandable in light of the experience of Judge Bruce M. Wright. One of the few blacks on the city's bench, Judge Wright claimed that his colleagues' practice of purposefully setting bail beyond the reach of defendants violated the Eighth Amendment and discriminated against blacks.

Known among police officers as "Turn 'em Loose Bruce," Wright was transferred from criminal court to civil court by the administrative judge in December 1974, after the police union and district attorneys waged a campaign against his practice of setting low bail for arrestees with long criminal histories.†

From the outset, Vera's operation of PTSA was viewed as a temporary arrangement designed to rescue the floundering ROR effort from the financially strapped Office of Probation. With substantial aid from LEAA, Vera was able to turn the program around; within two years it was able to reestablish its

---

*The controversy surrounding the agency's recommendation for the release of David Berkowitz—the famous "Son of Sam" murderer—is a case in point.

†This unprecedented transfer caused considerable protest. The Association of the Bar of the City of New York issued a report that called for Judge Wright's return to the criminal court. A suit on behalf of Judge Wright was filed by the Center for Constitutional Rights in Federal Court, S.D.N.Y., alleging unconstitutional improprieties on the part of the New York State judiciary. The suit was dismissed shortly after Judge Wright was returned to the criminal court on February 27, 1978. Wright later voluntarily returned to civil court, and the matter was closed. It should be noted that the defendant in his last controversial release decision was eventually acquitted of all charges.

reputation.* In its third year, PTSA, with Vera and city officials, developed another plan to institutionalize itself as a permanent city program. It created a public benefit corporation, the New York Criminal Justice Agency (CJA). Established in 1977, the agency provided services to the city on a contract basis, with funds supplied by LEAA and local taxes.

CJA was established in the midst of the city's worst financial crisis in recent memory. Even before it was fully launched, PTSA had to tighten its belt and reduce its commitments. This early retrenchment provided an opportunity for an unplanned, natural experiment to test the impact of the program. Faced with a dramatic reduction in funds, CJA made across-the-board cuts and experimented with alternative services in some boroughs. As a temporary measure in Queens during late 1976 and early 1977, the agency delayed interviews with arrestees until after arraignment, in order to avoid unnecessary interviews.

During this six-week period, 1,446 arrestees were arraigned in Queens Criminal Court. Of these, 378, or slightly more than 26 percent, had their cases disposed of at arraignment, either by pleas of guilty or by dismissals. Another group (48.5 percent) received RORs from the judges even though they had not been interviewed by the pretrial agency. Of those who remained in jail, a few were held without bail because they were wanted in other jurisdictions or were also charged with parole or probation violations, and the rest had bail set. Some were able to post bond and secured release on their own, but others were not.

How did the agency help this remaining group? One-third

*The Office of Probation maintains that the reason was that PTSA obtained substantial support from the LEAA, and that if it had received this amount, it, too, would have performed more effectively.

of them were never interviewed because they were not present at the courthouse at a time (usually nights and weekends) when a representative of the agency was on duty.* Of those arrestees who were interviewed, many were not recommended for ROR because of insufficient community ties or outstanding bench warrants (often issued for a past failure to appear). Only a handful were recommended for ROR, and eventually eleven of them were released on their own recognizance or on reduced bond.

How effective, then, was this intervention? Not very, according to CJA's report: "Adding the 11 PTSA-related releases to the 620 that were ROR'd or made bail on their own, the overall release rate [due to the Agency's intervention and recommendation] increases about one percentage point, from 58.05 percent to 59.08 percent."[24] This report states that the agency might have improved its operations had it employed a larger staff and maintained longer hours. Still, it concluded, even with more resources, PTSA impact would have been minimal.[25]

While the comparison is far from perfect, it is still useful to contrast the results of this "no early action" strategy in Queens with practices in Brooklyn, where the agency retained an active early interview strategy.[26] The comparison is revealing: only 41 percent of all arrestees in Brooklyn were released while 58 percent were released in Queens.[27] That is, despite a large, active program in Brooklyn and a smaller, passive program in Queens, the Queens courts released a much higher proportion of arrestees. The relative insignificance of the Queens intervention cannot be dismissed.

Finally, release rates during the Probation- and PTSA-run

*Much of the problem was due to transportation schedules to and from the courthouse and jail. Women are especially handicapped. They are often held in a central or regional facility, far removed from the court, their attorneys, and their families and friends.

periods indicate that the courts maintained roughly the same release rate—40 percent—providing still another indication of the marginality of the efforts with respect to affecting pretrial releases.

Another important claim of pretrial release programs is that they can lower the FTA rate. In Queens, the FTA rate for those released on their own recognizance by the court without the assistance of the pretrial release unit was 9.7 percent. This was *lower* than the FTA rate of a representative sample of ROR cases in Brooklyn under PTSA (13 percent FTA rate) and under the Office of Probation (23 percent). In light of these figures, the 9 percent rate for the Queens nonprogram is impressive. It is, in fact, equal to the FTA rate for those arrestees in Brooklyn who were designated "best risks" by PTSA.[28]

While one might regard the nonappearance of one out of every ten or eleven *apprehended* suspects as devastating, in fact most professionals find it tolerable. First, failure to appear does not necessarily mean flight; in fact, many of those who do not appear at one scheduled court date appear at the next. This has led many pretrial release specialists to distinguish between "passive" and "willful" FTAs; by all accounts, the latter group is very small. Second, those who fail to appear are usually charged with petty offenses, and while police and court officials regard nonappearance as a nuisance, they rarely enforce orders to appear, preferring to wait until the person is rearrested on another charge. In most cases, nonappearance is viewed as only slightly more serious than the failure to pay traffic fines. Some courts even encourage nonappearance. By setting low bonds for certain petty offenses and hoping for no-shows, these courts can quickly close out a case by ordering a bond forfeiture.[29]

While high FTA rates are often cited by judges as the reason for not liberalizing pretrial release, this argument masks their real concern, their reluctance to release arrestees charged with violent crimes. While many judges admit this in private, few do so publicly, preferring to hide behind high FTA rates and high bond. However, the telling evidence about official concern with FTAs is that police rarely serve FTA warrants and prosecutors rarely prosecute them.

While the magnitude of the impact of ROR programs continues to be debated, other services they provide are less controversial. In particular, in New York the Criminal Justice Agency has been instrumental in streamlining pretrial release practices and in contacting arrestees after their releases to remind them of their court dates. Unlike the earlier effort to place the pretrial release agency under probation, the institutionalization of CJA as a nonprofit corporation has been successful. As of this writing the agency has effectively served a number of functions without the problems that overcame the earlier pretrial release program administered by the probation department.

The California Model of Pretrial Release

Bail reform has developed differently in California. In New York, the ideal among reformers is the eventual elimination of bail bondsmen; in California and elsewhere, because of the power of bondsmen, nonmonetary conditions of release have received secondary consideration. The California legislature has mandated that each court develop a bail schedule (in essence, a price list) specifying the amount of bail required for each offense in the criminal code. Arrestees can be released as soon after arrest as practical, thereby avoiding the need to hold someone until arraignment, if they post the specified amount. This procedure does not preclude the possibility of release on recognizance or creation by local courts of special pretrial release units, and several counties, including Alameda (Oakland), San Diego, Los Angeles, and San Francisco, have created such agencies.

One key difference between California programs and those in New York City is that the former intervene *after* arraignment, after the opportunity for automatic release provided by the bail schedules has been invoked. Thus, to the extent that people who can raise money or who possess enough collateral to satisfy a bondsman are also those with strongest ties to the community, the most likely candidates for release on recognizance are those least likely to use it. Court-sponsored pretrial release units that promote ROR are clearly supplementary programs designed to handle those arrestees not able to secure release by money bail.

Because they are seen as supplementary to bail schedule-release, none of the programs in California is as large and elaborate as the program in New York City. In 1975, Los Angeles County, with a population of roughly 4 million, had a pretrial release budget exactly half that of Brooklyn, a borough with less than half its population.[30] Similarly, the budget for the San Francisco OR (own recognizance) unit in 1977 was $200,000, only slightly more than the budget for pretrial release operations in tiny Staten Island.[31] Of all the OR units in California, only the one in Oakland has had a budget, facilities, and staff that rival the per capita effort in New York City. Serving a population roughly the size of San Francisco, in 1977 the Oakland OR unit had an annual budget of slightly over $1 million, five times the amount allocated for San Francisco and slightly more per capita than for New York.

Bail Reform

The Oakland OR Unit

Supported by the Ford Foundation as part of its target cities program, the Oakland OR unit began operation in 1964. The agency interviewed likely candidates for OR release immediately before arraignment, but usually several days after arrest and after the automatic bail schedule had gone into effect. The OR program differed from the Vera model in other ways as well. It did not develop a community ties point scale but simply wrote narrative reports on defendants' ties to the community. The Oakland project was operated by the county-run probation department, and it was staffed with regular probation officers on rotating assignment, not with volunteers. And the program did not develop a follow-up system.

During its first two years (1964–66), the Oakland OR unit did not make a substantial dent in the number of pretrial detainees in the county jail. A comprehensive report of its operations for that period shows that the percentages of people released on recognizance rose only slightly from the previous (nonproject) year. Felony releases increased from 6 to 9 percent and nontraffic misdemeanors from 3 to 8 percent. Furthermore, most of these increases did not involve the real targets of the program, detained arrestees too poor to finance bond. Indeed, after two years, the number of people charged with misdemeanors who remained in custody until disposition actually increased.[32]

Although well intentioned, the unit's staff was so afraid of failure and bad publicity that it contributed to more conservative rather than more liberal pretrial release conditions. A good deal of the staff's time went into wasted efforts. Many arrestees the project considered to be good risks were able to post bond and secure their own release before the project staff could complete its investigation and ROR. Some of the probation staff did not bother to recommend ROR for those they thought would be convicted or sentenced to prison.[33] One careful study of the project's first two years concluded that "in considering the over-all problem of pretrial detention, the impact of the Oakland Project activity seems somewhat a 'drop in the bucket,' " and that those "seeking *major* changes in this phase of criminal justice will probably have to explore more heroic measures than those employed in the Oakland or similar projects."[34] On the other hand, the study did acknowledge that as a result of the project, for the first time release on recognizance was being used with some regularity, thus establishing a precedent.

The County Board of Supervisors refused to take over the program after the Ford Foundation's two-year grant ran out, and Oakland's first organized effort at establishing a pretrial release program withered for lack of local funding. It was not until five years later that an ROR unit was reestablished, this time as part of a court mandate to relieve overcrowding in the jails and offer release opportunities to those unable to afford to post bond.

One year after the court's action, in 1973, the county probation department obtained a two-year grant from LEAA to institute its pretrial release unit, and in 1975 the county agreed to continue funding the unit. Thus, ten years after the pilot project, local tax funds were finally being used to support the OR unit.

A comparison of pretrial release figures under the program and without it reveals some useful information.[35] Releases on recognizance declined after the reestablishment of the OR unit. Preproject releases averaged 20.4 percent, while project-recommended releases dropped to less than 17 percent. This decline was even greater for those charged with misdemeanors, presumably those most likely to be clients of the OR unit.[36]

Bail Reform

While RORs under the project declined, so, too, did the FTA rate, suggesting that the project concentrated on selecting only "good risks" at the expense of maximizing releases per se. The program's apparent conservatism might be attributed to criticisms that there were too many FTAs. But this is not a complete explanation. While FTA rates were reduced (from 32.6 percent to 30.3 percent), the change may have been the result of random variation and not of the project's efforts. Indeed, FTA rates did not decline in all branches of the county courts; in some they increased despite the decline of RORs.[37]

The explanation for the decline of RORs after the creation of the agency designed to promote them lies in what might be termed the paralyzing effect of information, or tyranny of incomplete information. The OR unit was required to obtain and verify more information than judges by themselves typically used. Recall that Paul Lazarsfeld's study of the community ties index used by New York City's PTSA showed that detailed indices were no more reliable than responses to one- or two-item questionnaires (and that none of them was very reliable). Lazarsfeld might have added that as the amount of required information increases, so, too, does the cost of gathering it and the likelihood that it will not be obtained and verified. (In the absence of complete and verified information, the Oakland OR unit remained silent, which in effect meant that it gave a negative release recommendation.)

The decision not to make recommendations in the absence of complete and verified information was premised in part on a desire to reduce error; yet the only type of error considered was that of recommending ROR for someone who subsequently would fail to appear. Errors caused by nondecisions—resulting in detention of those who would have appeared had they been released—were not considered. Thus the OR unit leaned in the direction of detention in order to minimize the likelihood of *identifiable* errors, even as the policy led to an increase in the undetectable errors of continued detention.

The pattern of the Oakland OR unit in the mid-1970s is consistent with what sociologist Forrest Dill found in his study of that same project in 1968. Its preoccupation with organizational survival led it to avoid "failure" at all costs. One sure way to do so was to minimize the number of ROR recommendations.[38] Dill's hypothesis is supported by a Lazar Institute report on the Baltimore pretrial release program, which concluded that the agency was unwilling to risk the adverse publicity that might accompany a slight increase in its FTA rate, even for the sake of a dramatic increase in the ROR rate.[39] The direct comparison is, of course, with the pretrial release unit in New York when it operated under the Office of Probation. There, too, the demand for more information on more people led to a decrease in RORs without achieving any additional benefits of another type.

During the mid-1960s, just before the Oakland project got started, the ROR rate hovered around 10 percent. Ten years later, during the period of its demise, the ROR rate was more than 30 percent. This shows that while there was a dramatic increase in the use of ROR during the decade, it was not directly attributable to the presence of a pretrial release program. On their own, judges were capable of sustaining a fairly high rate, and, indeed, the figures for 1975 and 1976 suggest that the special OR unit lowered the rate of those released on their own recognizance.

Other ROR Programs

San Francisco. The ROR program in San Francisco provides an especially useful comparison with the program in neighboring Oakland. In 1976, Oakland's office had a budget well over $1 million, was supported by county tax funds, and was staffed by trained probation officers. It used an extensive interview form and provided follow-up notification services. In contrast, San Francisco's program operated out of a cubbyhole in the county courthouse, relied heavily on volunteers, and had a small, low-paid, nonprofessional staff. It was a shoestring operation, supported by a variety of private sources, and had a budget roughly one-fifth that of Oakland. The Oakland program had an elaborate computerized information system; the San Francisco office kept its records on three-by-five cards filed in old shoe boxes. Yet, despite these differences, there is no demonstrable evidence that Oakland's pretrial release and FTA rates were any better or worse than San Francisco's.

Washington, D.C. Since 1966 the courts in Washington, D.C., have been able to avail themselves of both the Federal Bail Reform Act, one of the nation's most liberal pieces of pretrial release legislation, and the services of the D.C. Bail Agency, one of the largest and most well-funded pretrial release programs in the country. These are usually given as reasons for the low pretrial detention rates in Washington. For instance, Wayne H. Thomas, Jr., notes that in 1962, Washington detained until disposition fully 62 percent of all those charged with felonies, releasing none of them on their own recognizance; but by 1972, the percentage of detainees dropped to 31 percent, with 56 percent of felony defendants released.[40] Examination of the history of bail reform in Washington suggests that the great changes in release practices (and in the federal courts generally), occurred *before* the bail agency and the act took effect. Thomas himself notes that in 1965 the pretrial release rate for *felony* defendants was 58 percent, and that many defendants were released on their own recognizance.[41] Thus, two-thirds of the reduction in detention that Thomas appears to attribute to the D.C. Bail Agency and the act took place prior to their creation.[42]

Yale law professor Daniel J. Freed was in charge of the bail reform effort in the Department of Justice in the mid-1960s. Though he defends the importance of both the D.C. Bail Agency, which he helped to create, and the Bail Reform Act, which he helped to write, he acknowledges their limited impact:

> The biggest effect on pretrial release in the federal system did not occur as a result of the Federal Bail Reform Act, but earlier—in response to one piece of paper. It was an order by [Attorney General] Robert Kennedy, ordering U.S. Attorneys to accept ROR for defendants in cases where there was a low likelihood of the defendant's fleeing the court's jurisdiction. This single action decreased pretrial detention by 75 percent in the federal system.[43]

Other Cities. Using time-series analyses and quasi-experimental methods, Roy Flemming, C. W. Kohfeld, and Thomas Uhlman studied a pretrial release court in "Metro City" (a large northeastern city).[44] Their findings support the evidence presented earlier. Tracing the release rates before and after introduction of the reform, these researchers found that "creation of the pretrial services agency only marginally affected recognizance release rates," and concluded that "the slight improvement in recognizance release rates after the creation of the pretrial services agency could have occurred by chance."[45]

In a study of Baltimore and Detroit, Roy Flemming found similar patterns.[46] In Baltimore, creation of a pretrial release agency had little effect on the rate or type of pretrial release. In Detroit, despite lack of a formal agency, he found substantial liberalization of conditions for and rate of pretrial release. He attributes Detroit's changes to increased politicization of the bar and especially of the relatively large and well-organized black defense bar. (In another study, Isaac Balbus concluded that accused rioters fared better in Detroit than in other riot-torn cities, thanks to the politically conscious and active black defense bar in that city.)[47] Flemming's findings suggest that the political power and organization of supporters are more important than formal changes. Finally, a multicity study by Mary A. Toborg and Martin D. Sorin for the Lazar Institute revealed findings of little or no program effectiveness as indicated by a number of criteria including securing court appearances, identifying dangerous offenders, and overall cost effectiveness.[48]

Cities with No Programs

So far, the examination of pretrial release agencies has focused on before-and-after comparisons or on comparisons among different types of programs. Another way to approach the problem is to make comparisons of jurisdictions with and without specialized pretrial release programs. The expectation that cities with programs would have higher release rates than those without programs is not borne out by available data.

A detailed study of Rhode Island—a state without any program—revealed that from 50 to 55 percent of all felony defendants, including those charged with capital offenses, were released on their own recognizance.[49] Of all the cities with pretrial release programs studied by Wayne Thomas, only one, Washington, D.C., equaled the rate in Rhode Island; most cities did not even come close.[50] Similarly, a study of bail practices in Cleveland found that 68 percent of all felony defendants were eventually released on bail or on their own recognizance, despite the absence of a special pretrial release program at that time.[51] Studies of Milwaukee and Richmond, Virginia, which at the time had no special pretrial release units, revealed release records comparable to those cities with large and active programs.[52] Finally, Roy Flemming found that Baltimore, with a pretrial release agency, had a higher detention rate than Chicago and Detroit, without agencies.[53]

Assessment of the Pretrial Release Movement

Although the goals of Arthur Beeley, Caleb Foote, Herbert Sturz, Daniel Freed, and Robert F. Kennedy have not been fully realized, we have witnessed a dramatic liberalization in pretrial release in recent years. The changes are generally attributed to special pretrial release units and their use of techniques to predict likelihood of flight. However, close inspection of these programs reveals something of the contrary, and evidence suggests that change can take place in jurisdictions without agencies or where agencies do not intervene. At best, special agencies play only a marginal role in pretrial release. This does not mean that they cannot be important, only that their effectiveness has not yet been demonstrated.

Despite their questionable value, there are several reasons why specialized pretrial release agencies have been given such prominence. First, they have generally been welcomed by court officials, but perhaps for the wrong reasons. Because the problems they are aimed at have not been fully diagnosed—or, more properly, there has been a fundamental conflict of values in the diagnosis—their acceptance has more to do with the game of symbolic politics than with demonstrated commitment to change. Judges and prosecutors have a common interest in trying to avoid responsibility for detaining people who cannot afford to post bond and for released defendants who subsequently fail to appear in court. Traditionally, judges have claimed that they were guided by prosecutors' recommendations, while prosecutors have claimed that the ultimate authority has rested with judges. Now both can point to the pretrial release agency for release problems. The agency is a convenient scapegoat and can contribute to the further fragmentation of an already overfragmented system.

A second factor in the inflated importance of these agencies is an uncritical acceptance of the veneer of science and rationality. Pretrial release programs claim that they can scientifically predict which defendants are likely to appear in court. There is a library of research in this area, correlating court appearance rates to factors ranging from the number of siblings a defendant has to the promptness with which he pays his utility bills. Numerous manuals have been written on the construction and interpretation of point scale systems, but even the most objective scale allows for considerable subjective judgment. Further, pretrial release units only make *recommendations* and do not automatically release all high-scoring defendants. This makes it more difficult to determine accurately how reliable and valid the objective predictors are. Finally, those programs that have developed the most comprehensive predictor system also provide for elaborate follow-up. Thus, it is impossible to determine whether a low FTA rate is due to the predictive accuracy of the index, to follow-up, or to both.[54] Follow-up seems to be the more effective feature of release agencies.

We have seen that one of the most well-regarded innovations in the criminal courts was initiated, implemented, and at times institutionalized without much careful assessment of the accuracy of its technology or the nature of its impact. It should be a matter of considerable concern that some of the most obvious and basic issues were not subjected to even cursory research until fifteen years after the first experiment and research agenda were established, and long after dozens of programs based upon questionable assumptions had been developed. This was possible because the new release units performed elusive tasks of such low visibility that few were in a position to speak knowledgeably about them. And in a fragmented system, fewer still were in a position to challenge their claims.

I do not mean that the bail reform movement has had no effects. Clearly it has had a major and significant impact. Nor am I arguing that these agencies are redundant. Increasingly, those closest to them recognize their limitations and seek to justify them in terms of a multiplicity of functions. Thus, like so many other areas of the criminal process, here, too, official function and informal task diverge and in so doing make discourse on policy and impact difficult.

One of those who has cut through the formalism surrounding so much discussion of pretrial release is Bruce Beaudin, director of the D.C. Bail Agency. When queried about some of the issues raised above, he acknowledged that "there are no reliable predictors of future appearance" and that the search for predictors is, in effect, an attempt to "con the judges and public by developing an elaborate hocus pocus which gives an appearance that our decisions are objective and based on good evidence and the like. . . . Basically . . . we're simply trying to get as many people released as possible." When asked why this charade was necessary, he responded with candor: "You don't need a bail agency if the judges [are] doing what they are supposed to be doing. . . . You need a bail agency to keep the system honest, to get others to do what they ought to be doing!" But, he continued, this function was most needed when the idea of ROR was new. He argued, "The people who are not now released in D.C. are parole violators. Almost everyone else is released and they would be released even if we were not there."[55]

The limited impact of the agencies in Oakland, Queens, Brooklyn, Washington, and New Haven supports his conclusions. Indeed, the experiences of those cities without agencies—Cleveland, Richmond, Milwaukee, greater Providence, and Washington, D.C., prior to 1966—suggest that these agencies may not even be necessary for the limited function he suggests they serve.

Still, I cannot conclude that these programs have served no purpose. Clearly organized follow-up to remind people of pending court dates reduces the FTA rate. This may be reason enough to justify pretrial release units, although the same results might be obtained if the secretaries to defense attorneys took the time to telephone a reminder to their clients, and if prosecutors and judges occasionally prosecuted and convicted no-shows.

More important, a pretrial release program may reduce discrimination by court officials. At the time of this study, Rhode Island had one of the highest pretrial release (ROR) rates in the country, yet had no special release program. Still it had another and serious pretrial release problem: blacks were at a noticeable disadvantage in obtaining ROR,[56] and some judges were exceedingly reluctant to grant ROR. A public monitoring agency with standardized criteria would discourage such outrageous discriminatory practices as these. Similarly, Flemming found that in Baltimore, although the agency did not measurably increase the rate of ROR, it did affect who got released and as such reduced arbitrariness and racial discrimination. And Toborg and Sorin found that ROR programs reduced the disparity of treatment of Spanish-speaking arrestees.

One last positive feature of pretrial release agencies must be acknowledged. Their creation and existence have been effective in calling attention to the problems of the accused. Collectively, these agencies have mounted a loud and effective campaign for adoption of release alternatives less onerous than money bail. Herbert Sturz's early modest experiment generated the enthusiasm and concern necessary to launch a nationwide movement. His enterprise has generated a permanent interest group, complete with its own national organization headquartered in Washington, which effectively monitors practices and works for expanded pretrial release procedures. Concern with pretrial release is now institutionalized in a way it never was in the 1920s.

Institutionalization and Permanent Funding. Pretrial release programs lead precarious lives. Once initial sources of "free" or "outside" funds are exhausted, the agencies are left to fight for far fewer, and more difficult to obtain, local tax funds. This has caused some agencies to lead on-again/off-again existences, others to experience dramatic expansions and contractions, and all to worry constantly about survival.

Forrest Dill has concluded that the preoccupation with funding has led pretrial release agencies to all but abandon their original goal of reducing pretrial detention.[57] Some energy is diverted for purposes of financial survival, and still more is used up in reaching accommodations with court-related organizations. Adopting positions acceptable to prosecutors, judges, police, and probation departments often means significantly softening original commitments to pretrial release. Agencies' instability and marginal status in an adversary process lead them to become vigorous players in the bureaucratic funding game at the expense of their original mission.

Presumably, if the court were doing its job properly, pretrial release agencies would be unnecessary. Their very existence, then, is an implicit criticism of the courts. This leads to a tendency to steer clear of court authorities who, by design or inertia, have contributed to the target problem. In many larger cities, there is also a desire to avoid an entrenched and lethargic civil service. High educational requirements, entrance tests, seniority systems, elaborate promotion policies, rules about hours, benefits, and the like are all resisted by program advocates who prefer to cut through red tape and serve the "real" needs of the accused. But this circumvention of established institutions means that these programs are always on the fringe of power and security, a fact that makes them vulnerable to drastic budget cuts and high turnover, in short, institutional instability.

Evaluation. Those who promote pretrial release programs *know* they work; they are morally committed to them. Many program evaluations are in effect little more than promotional pieces and certainly are not hard hearted and skeptical assessments. On the other hand, critics often have little incentive to challenge these so-called evaluations.

Programs can claim great success while having only marginal impact. They can provide impressive records to funding agencies, which are not close enough to the subject to challenge evidence.

At the time of this study, there has not been one scientifically acceptable evaluation of any major pretrial release program. Although enough reports have been written to fill a small library, few meet even elementary research standards. Many of these seemingly impressive data-based reports are little more than promotion pieces that ignore or beg the hard questions.[58] Particularly disappointing is the fact that many of these programs have been funded by LEAA as "experiments."[59]

Even the major surveys of pretrial release efforts commissioned by the National Science Foundation did not always ask the right questions.[60] The NSF-supported researchers focused on differences of administration and organization among release agencies, ignoring the question of whether any type of agency at all was necessary. More recent research has sought to overcome these problems and use imaginative research designs to examine the nature and effects of pretrial release programs. This research, too, suggests a minimal impact of such programs.[61]

Future Strategy. There are many indications that pretrial release can be liberalized without specialized agencies and large staffs. In a number of cities in recent years, jail overcrowding has forced judges to take immediate action, without waiting for the creation of such agencies. In response to federal court orders limiting the size of jail populations, judges in Baltimore, Maryland, and Santa Cruz, California, were forced to quickly adopt liberal pretrial release policies, and reports on these actions indicate no significant changes in rearrest or FTA rates.[62] Judges in New York City followed this same course of action in the fall of 1970, when rebellions in the Tombs forced them to release arrestees prior to trial. In the late 1960s, the Detroit bench and bar responded to that city's riots with liberalization of pretrial release and achieved notable successes.

Citation arrest and summonses in lieu of arrest at the site of the incident or at the station house are now common in many cities. These innovations were brought about in large part by the police themselves to reduce the amount of time they spent booking people on nonviolent charges and waiting for arraignment in court. Officers often have ready access to information about the accused's ties to the community and thus may be as competent as any specialized agency to make informed decisions about likelihood of court appearance.[63]

Police are capable of making fair and effective release decisions in the absence of both judicial supervision and a specialized agency. Experience in New Haven is a case in point. There, despite the existence of a special pretrial release commission, police released almost all arrestees long before the commissioners had an opportunity to intervene, and they were quite liberal in granting ROR.

Advocates of pretrial release programs usually avoid the issue of whether their elaborate information systems and predictive techniques should be used for purposes of preventive detention. Some pretrial release programs are already providing judges with estimates of likely dangerousness, and some are now trying to develop predictors of future dangerousness as an extension of their research on predictors of future court appearance. There is little reason to believe that the effort to predict dangerousness will be any more successful than the effort to predict failures to appear. There is, however, great reason to believe that the facade of science will be used to justify increased harsh treatment of allegedly dangerous people. This may be one of the lasting, if unwitting, consequences of the administrative and scientific approach to bail reform.

While few pretrial release units formally embrace the concept of preventive detention—and many on their staffs vehemently oppose the idea—increasingly, agencies are collecting statistics about the numbers of their releases rearrested on "bail crimes" and are coming to accept preventive detention. In the future, these figures may also play a role in their recommendations. A number of states are now in the process of rewriting their pretrial release statutes to permit broad authority to preventively detain, and the District of Columbia has had statutory provisions for preventively detaining without bail since 1969.

Although the District's provision has all but lain dormant since its adoption (prosecutors find it more convenient to ask for high bail than to go through the complicated procedures required for preventive detention), it does specify in great detail those factors thought to be associated with future dangerousness. Many states are considering similar provisions. However, a study conducted at the Harvard Law School presents strong and convincing evidence that the factors enumerated in the D.C. statute are not accurate predictors of dangerousness.[64] The study found that the detention rate in Boston would have more than tripled if the act had been applied there. More important, of those Boston releases who would have qualified as "dangerous" under the terms of the statute, very few were rearrested on serious charges while they were waiting for trial. For every so-called dangerous person who was rearrested during this period, there were from seven to fifteen others (depending on various definitions used in the study) also meeting the "dangerous" criteria who were not rearrested. Many others who did not meet the criteria were rearrested during this period. In short, the predictors erred in both directions. The study concluded that the "dangerous" criteria were in essence arbitrary and invalid. More recent research on this issue arrives at similar conclusions.[65]

Given the errors in scientific prediction, there is much to be said for pursuing an "unscientific" rights strategy in bail reform. Alexander Bickel spoke eloquently about the positive functions of ambiguity and the social benefits of not always trying to reconcile conflicting principles.[66] These firmly held and deeply felt interests—concern for the rights of the accused and the right to bail, on the one hand, and concern for the safety of the community, on the other—may not be reconcilable. If so, any choice will be wanting, and the best one can hope for is a partial solution. Under such conditions, I argue for the limited goal of securing appearance at trial, even if from time to time it permits the hypocrisy of setting bail beyond reach. Despite abuse, this policy has the virtue of formally incorporating the aspirations and ideals central to the criminal process, the presumption of innocence. Although it claims the virtue of honesty, a policy permitting preventive detention compromises this principle, and it, too, is subject to abuses of its own. That there are no commonly recognized and reliable predictors of dangerousness makes abuse all the more probable. When formal discourse gives way to functional analysis, and principles are understood in the context and conditions under which they operate, what at first appears to be second best may emerge as preferred. So it is with the policies of bail and preventive detention.

Reliance on a strategy of rights also reinforces a central role of the courts, that of allowing challenges by interested parties. In recent years, the judicial process—both prosecution and defense—has been dramatically improved by strengthening both the adversary system and the rights of the accused. There is reason to believe that a rights strategy for pretrial release based on bail would have the same salutary effect.

Professor Caleb Foote's hope that a crisis in bail would be resolved through constitutional law reform has not been fulfilled. The Supreme Court has held that bail set beyond the reach of the accused is not necessarily excessive.[67] Other legal challenges to current bail practices have met equally numbing defeats.* Perhaps these have been conditioned by the conservative tenor of the Burger Court, but they may also be due to a sense of complacency about the administrative approach to bail reform as reflected in pretrial release programs.

It is surprising that the appellate courts have not expanded the right to counsel to the earliest stages of the criminal process

---

*One of the most creative of these suits, *Wallace* v. *Kern,* 520 F. 2d 400 (2d. Circuit, 1975), unsuccessfully attempted to establish due process standards in bail hearings in Brooklyn, New York. The sympathetic decision by the district court was reversed by the appellate courts, which held that interference with state bail practices violates the doctrine of federalism.

and insisted on a *meaningful* bail hearing. It remains the rare exception that a defendant will have consulted with an attorney before the first decision to consider pretrial release is made.[68] Though counsel is available later to make bail motions, judges are notoriously loath to reverse an initial bail decision. There is clearly a critical role for assertive defense counsel in the release determination process. Pretrial release is not a neutral, value-free decision, any more than adjudication or sentencing are. One's constitutional sensibilities would be offended if defense counsel did not interject in a sentencing hearing; yet there is no general concern when such passiveness attaches to bail decisions. In both instances, there is a need for the careful scrutiny and questioning that only an advocate is likely to give; information may be incomplete, inaccurate, out of date, ignored, carelessly interpreted, and so on, even when marshaled by a special agency.[69] This suggests the value of early availability of counsel, although a rights strategy, if effective, could serve to clarify principles so that they could be executed by police and jailers at early stages of the criminal process.

A good example of an alternative approach to bail reform through an advocacy strategy is the special defender service of the criminal defense division of the New York City Legal Aid Society. The special defender service assists an attorney in preparing bail reports. It gathers information on the defendant's background and social history and provides the court with alternatives in pretrial detention. At the time of this study the program's resources and staff were extremely limited, and it was mainly used for presentence reports to encourage the court to consider alternatives to incarceration. Nevertheless, its advocacy outlook marks a decisive break with a neutral, administrative approach to bail reform.

Bail Reform

But where will already overworked public defenders obtain the necessary resources to convert bail proceedings into a rights-oriented advocacy process? Perhaps from pretrial release agencies, which arose in response to a real problem but now—at least in some places—are part of that problem.

Chapter 3

PRETRIAL DIVERSION

FOR the criminal courts, the 1970s was the decade of pretrial diversion. The idea was simple and appealing: social problems not appropriately or effectively handled by the courts were to be "diverted" to other agencies. The President's Commission on Law Enforcement, which proposed pretrial diversion in 1967, found that many criminal cases involve people with chronic mental or alcohol-related problems. In such situations, the commission argued, "it is more fruitful to discuss, not who can be tried and convicted as a matter of law, but how the offices of the administration of criminal justice should deal with people who present special needs and problems." The solution of the commission was the "early identification and diversion to other community resources of those offenders in need of treatment, for whom full criminal disposition does not appear required."[1] Diversion to social service agencies stood to benefit not only the accused but also everyone involved: it offered specialized services to those in need of them, freed the hard-pressed police and courts to be deployed more effectively elsewhere, and benefited the public at large by reducing recidivism.

Diversion was to operate as follows: after arrest, but before appearance in court, defendants with obvious psychological problems were to be referred to mental health facilities; those with alcohol-related problems were to be sent to shelters or halfway houses. The immediate popularity of diversion was due to the fact that, in the words of two knowledgeable observers, it offered "the promise of the best of all worlds: cost savings, rehabilitation, and more humane treatment."[2]

While the President's Commission originally proposed diversion for those with obvious mental or alcohol-related problems, the concept was quickly broadened to cover a host of others—first offenders, the economically disadvantaged, and the unemployed. "Treatment" was expanded to include group therapy, job training, job placement, and welfare referral.

This expanded notion of diversion found theoretical support in labeling theory, a sociological perspective holding that deviance is not a quality inherent in the individual—the so-called deviant—but is a consequence of the social process of being identified or labeled as a deviant. Deviance, according to this view, is a condition "created" by society.[3]

While diversion appealed to proponents of rehabilitation, it was also accepted by hard-nosed judges and prosecutors, people not likely to be sympathetic to labeling theory. Faced with a chronic shortage of resources and an increasing case load, they welcomed an alternative, more rapid means of processing petty offenses. And since proposals for diversion invariably permitted prosecutors the final word on admittance, few feared any threat to their authority.

The Rapid Rise of Diversion

Informal diversion has long been practiced in American courts. Indeed, one of the functions of a good defense attorney is to try to get the prosecutor or judge to drop the charges if the accused makes some effort to overcome his problem or make amends to his victim. However, special diversion programs are new.

In 1967, the Vera Institute in New York City and the National Committee for Children and Youth in the District of Columbia asked the Department of Labor for funds to support their proposed diversion programs. Both organizations had received prior Labor Department funds, authorized in the Manpower Development and Training Act of 1962, to support vocational education and work release programs for sentenced prisoners. Their theory, which was accepted by the Department of Labor, was that criminal activity is motivated in part by unemployment, and that by supplying an arrestee with a job in lieu of conviction and sentence, diversion could "reduce the likelihood of future criminal behavior, reduce the work of the courts, and provide a tax paying citizen."[4]

Both Vera's Manhattan Court Employment Project and Washington's Operation Crossroads, as the projects came to be known, began operations in early 1968 and were immediately hailed as bold and successful new experiments. In 1971, the Department of Labor sponsored a second round of diversion programs, this time in nine cities (Atlanta, Baltimore, Boston, Cleveland, Minneapolis, San Antonio, and three sites in the Northern California Bay area: Hayward, San Jose, and Santa Rosa). Billed as eighteen-month "experimental" tests of different diversion "strategies," they cost the department $3.5 million. Despite earlier evaluations of the New York and Washington projects, which found that "neither project was able to clearly substantiate the rehabilitative impact of intervention,"[5] the department proceeded to support these new projects, confident that they could overcome the shortcomings of the initial programs and test hypotheses about successful intervention among a "broader range of defendant groups across several state areas."[6] An elaborate and expensive evaluation component was included in plans for these programs, and funding was conditioned upon local acceptance of it.

In spite of these careful plans and good intentions, once the local projects had received their funds, they refused to cooperate with the evaluators, failing to maintain records, select control groups, and collect data according to the provisions of their agreement.[7] And having completed the arduous administrative work required to fund so diverse a group of programs, Washington-based DOL officials had neither the inclination nor the proximity to monitor every detail in the agreements. The result was that the time-consuming and expensive evaluation report revealed little about the projects.

Nevertheless, the Labor Department's programs received a great deal of favorable publicity. Since its establishment in 1968, LEAA had been severely criticized as being overly enamored of police hardware and insensitive to the other needs of the criminal justice system. Diversion emerged as a program by which LEAA could redeem itself.

Embracing the Labor Department's claims about the benefits of diversion (and ignoring the negative findings of early research), LEAA all but turned diversion into a household word. It revised its annual guidelines to require that states receiving federal funds had to initiate plans for diversion programs; it earmarked substantial portions of its discretionary funds for diversion programs; and it convened numerous national conferences and underwrote the cost of many publications extolling the virtues of diversion. Both Congress and numerous state legislatures drafted legislation providing for pretrial diversion and delayed prosecution, even though such statutes were often unnecessary since prosecutors already had authority to drop, delay, or alter charges.

Throughout the early and mid-1970s, LEAA continued to promote diversion heavily, and dozens of new programs— usually federally funded—were established across the country. According to Madeleine Crohn, former director of the Pretrial Services Resource Center in Washington, the number of pretrial diversion agencies grew from the initial 2 in 1967 to well over 200 by 1977.[8] The total annual price tag for these programs cannot be determined, although it ran into tens of millions of dollars.

But a decade after diversion first caught the imagination of reformers, interest in it began to wane. Since 1977, support for projects has declined, as LEAA first changed its priorities and later experienced its own cutbacks and loss of power.

Few local jurisdictions have been willing to absorb the cost of these programs, many of which had overhired staff and made preposterous claims of success while they were supported by federal funds. As a consequence, diversion programs were cut back—or dramatically shifted focus—about as quickly as they had sprouted up.

We have, then, something like a life cycle of an innovation. Fortunately, diversion has left a trail strewn with discarded reports and evaluations from which we can piece together its goals, successes, failures, and impact on the courts.

The Manhattan Court Employment Project: The First Big Step

When the idea of pretrial diversion first surfaced in the mid-1960s, the Vera Foundation persuaded the Department of Labor to expand its commitment to the criminally accused, and in 1968 Vera began its diversion program, the New York

Pretrial Diversion

Court Employment Project (CEP), under a $300,000 annual grant.

An early CEP document describes its objectives as follows:

> The Project's ability to convert a defendant's arrest from a losing to a winning experience benefits the defendant, the courts and the community; successful participants have their charges dismissed and leave the Project employed or in vocational or academic training; the overburdened criminal system is freed to attend to more serious cases since Project participants do not spend time in the overcrowded detention facilities, and successful participants make fewer court appearances and are less likely to be rearrested than the average defendant; and the community gains because individuals who may have been developing a lifelong pattern of criminal behavior are now on their way to becoming productive, taxpaying members of society.[9]

From 1968 to 1971, CEP operated in Manhattan, with a professional staff of twelve, which offered a mixture of counseling and job placement services. According to an early description, CEP's screening process was as follows:[10] to obtain clients, CEP would station one of its staff at the arraignment court. This person would approach potentially eligible arrestees, explain the goals of the project to them, and then invite their participation. If an interested arrestee met CEP's eligibility requirements—and if the prosecutor agreed—he would be released on his own recognizance, his case would be postponed for ninety days, and he would work with the CEP staff to develop an appropriate program of social services. If the participant met his obligations to the diversion program, CEP would return to court after the ninety-day period and recommend that charges be dropped. In most instances, CEP reported, the prosecutor and judge would accept this recommendation.

During its first three years, CEP focused primarily on job

training and placement. According to its own published reports, it was an immediate and overwhelming success. CEP reported that it was able to obtain steady employment for almost all of its participants and to more than double their incomes.[11] CEP also claimed a substantial reduction in rearrest rates for participants. A twelve-month follow-up study of rearrest rates among participants and a control group claimed that twice as many nondivertees as divertees were rearrested. This, the report argued, was proof that CEP "effectively reduced the incidence of rearrest among dismissed participants."[12]

After Labor Department funding was exhausted at the end of 1970, CEP entered into a contract with the city's Human Resources Administration (HRA) to provide services to the criminal courts. Between 1971 and 1975, CEP obtained increasingly larger contracts from HRA and expanded its services into four of the five boroughs. By 1975, it had an annual case load of more than three thousand and an annual budget of more than $3 million.

Despite the apparent success of its job-oriented program, once Labor Department funds were exhausted, CEP abruptly changed direction. With HRA support, beginning in 1971, CEP experimented with a variety of counseling and social service programs: individual therapy, group therapy, and instruction in birth control, remedial English, dressing for interviews, and the like. The move from the original direct employment goal was explained by the staff as the result of a gradual shift to younger clients "who need remedial training more than jobs" and a downturn in the job market. The CEP staff also pointed to recent positive evaluations. For instance, a 1977 study reported that CEP had a "55 percent success rate," a rate apparently high enough to satisfy CEP officials.[13]

Although CEP, HRA, and court officials were relatively satisfied with CEP, the city's financial crisis of 1976 brought its operations to a virtual standstill. CEP was forced to lay off all but a handful of its staff, and those remaining spent their time trying to locate new sources of funds. In late 1976, they finally badgered the city into renewing CEP's contract, but on a much reduced scale. When CEP reopened in 1977, its budget, staff, and clientele were roughly one-third what they had been two years earlier.

To some CEP officials, this cutback was a blessing in disguise. The reductions led CEP to concentrate only on those charged with felonies, and in so doing it sought to overcome the persistent criticism that it was little more than a dumping ground for prosecutors who did not want to drop weak cases. During 1977 and 1978, CEP handled about 1,000 cases per year and, at least in the beginning, maintained that it was successful.[14]

However, during the same period, the Vera Institute completed a major evaluation of CEP that showed that it was not significantly affecting case outcomes, and that despite the program's formal independence, the prosecutor controlled both access to and outcomes of the diversion process.[15] In short, the evaluation concluded, "independence was illusory."[16]

CEP leadership and its board of trustees wrestled with these findings, and in late 1978 came to the conclusion that pretrial diversion had been based on false premises, and that CEP should cease existence in its present form. Demonstrating a candor rare for institutional leadership of any sort, CEP's board voted that the agency "shall herewith not accept clients from the criminal justice system on a formal pretrial diversion basis as it has, historically, done."[17]

On the other hand, CEP leadership was impressed with its staff and remained convinced that it had an important mission. It voted to continue operations. CEP initiated a major shift: acknowledging its prior control by prosecutors, it sought to develop closer ties with defense attorneys and to become a new option in defense strategy. Once defense attorneys have identified clients likely to be convicted and possibly sentenced to jail, they can turn to CEP, which will tailor a counseling or community service program for their clients. The defense attorney can then approach the prosecutor and judge with this plan, arguing that his or her client as well as the interests of justice are best served by participation in the program.

While it is too early to assess this new approach, it appears to have overcome one major *structural* problem that has plagued so many pretrial diversion programs—domination by prosecutors. The strategy of the new institution is now shaped by defense attorneys, those people best able to make a judgment about the likelihood of incarceration. This revised successor to CEP has sought to accommodate itself to the adversarial system rather than trying, as in the past, to circumvent it.

CEP's Impact

One of the initial claims of CEP was that early diversion of "inappropriate criminal cases would allow the courts more time and resources to deal with more serious criminal matters." Yet diversion programs handle too few cases to have any impact on the courts. In its busiest year, 1975, CEP handled less than 2 percent of all the city's misdemeanor arrestees.

Although the CEP staff, like diversion staffs in other cities, felt that this low rate of participation was due to rigid eligibility requirements set by prosecutors, in actuality other factors were more important. For example, one study found that 44 percent of all eligible arrestees in New York who did not participate in the program rejected it either by themselves or after advice of counsel.[18] Another, reporting on CEP's activities in fiscal

Pretrial Diversion

1974–75, found that more than 50 percent of potentially eligible defendants preferred to take their chances with the regular path through the court.[19] Reasons for lack of interest varied: some felt diversion offered no real benefits; others were certain that charges against them would be dropped outright; and still others preferred to plead guilty and "get it over with" quickly. Ultimately, these reasons converge into one: diversion did not offer them very much. The relatively quick and painless trip through court was preferable to the long delay and the frequent meetings imposed on them by CEP, even if CEP meant they could avoid a criminal record.

Although CEP retained its original name until its effective demise as a diversion program in 1979, it ceased its direct involvement in securing employment in 1971, when it began concentrating on counseling and employment referral. If "employment leading to self-sufficiency and a reduction of recidivism" was once CEP's goal, other goals later replaced it. Perhaps it was easier to "rap" with clients in an office than to pound the pavement lining up employers willing to take a chance with someone in trouble. One careful study of CEP concluded that counselors "hope that [their] services will produce some long-term benefit to the client, but they express little conviction that it will."[20] Given this belief, it is not surprising that "success" in the program came to be defined as regular appearance for scheduled counseling sessions.

But even this minimal criterion was often softened. Counselors varied in their willingness to tolerate missed appointments; and success or failure, and with them dismissal or conviction, often came to depend on a counselor's tolerance. So when CEP reported that in 1977 it had a 55 percent success rate for the 532 people who entered the program, it meant only that 55 percent of the participants appeared more or less regularly at scheduled meetings.[21] This Orwellian definition of "success" is a far cry from the early claims that CEP secured employment for virtually all its clients who had been underemployed or only occasionally employed prior to entering the program, that a high percentage of these people reported being employed fourteen months later, and that their incomes had more than doubled.[22]

The Broader Significance of CEP

Until confronted with the results of the Vera evaluation in late 1978, CEP had maintained that it was successful. Beginning in 1969, CEP produced a series of favorable quarterly and annual reports, and advocates who previously had supported the idea on faith had hard data as reinforcement. In particular, CEP's 1972 reports came to play a major role in shaping the future of other diversion programs. After hearing of CEP's successes, Congress adopted legislation authorizing diversion programs in the federal courts,[23] and New York City's HRA agreed to appropriate more than $3 million per year for an expanded diversion program. CEP's figures also impressed LEAA, which immediately required states to make provisions for pretrial diversion programs as a condition for receiving federal funds. As a consequence, CEP quickly emerged as the standard against which other programs were compared. Perhaps more than any other research, CEP's early studies purporting to show dramatic success legitimized the idea of diversion.

A few skeptical voices were heard amidst this chorus of praise. In 1973, Harvard scholars Elizabeth Vorenberg and James Vorenberg published an article raising a number of serious questions about pretrial diversion and expressing concern that however well intentioned it was in theory, in practice it permitted the state considerable control over the lives of people not convicted of any offense.[24] Another voice in the wilderness, Paul Nejelski, examined diversion programs for juveniles and warned that pretrial diversion had the potential for expanding social control even as it claimed to contract it.[25]

Professors Norval Morris and Franklin Zimring of the University of Chicago voiced many of these same concerns, and Zimring went on to challenge the earlier claims of success. He found that CEP's widely cited studies were seriously flawed. Reanalyzing CEP's data and presenting some of his own, Zimring argued that contrary to its claims, CEP had not had any measurable success in reducing recidivism. He warned that these claims of dramatic successes had led supporters to expect far too much from diversion, and ended with a thoroughly disillusioning conclusion: "Our reevaluation of CEP tended to show that when defendants are diverted from the ordinary process of the criminal justice system, they do no worse than those who are fully processed through the police, courts, and correctional systems in New York City."[26]

Despite these warnings, enthusiasm for diversion continued unabated. By the time the critical articles had reached their audience, the pretrial diversion movement was in full bloom, and these warnings were swept aside. The initial report on CEP success had preceded them by a year or two and had become the standard evidence in support of diversion. The belief that diversion provided considerable benefits to the accused was steadfastly maintained, even in the face of mounting evidence to the contrary.

Still, the criticisms had impact. In response to Zimring's challenge and to growing skepticism about pretrial diversion programs, LEAA's research arm, the National Institute of Law Enforcement and Criminal Justice (NILECJ) earmarked more than $250,000 to underwrite a methodologically sound evaluation of pretrial diversion. Because of its premier status among diversion programs and its close connection with the Vera Institute, CEP was selected as the test program. Following Zimring's advice, the Institute-sponsored study compared the behavior of two groups of equally eligible and interested arrestees. By random assignment, one group was admitted to the program; the other was not. The resulting study, directed by Sally Baker and Susan Saad of the Vera Institute, is a model of evaluation research—well designed, well executed, and well presented.[27]

Baker and Saad found that while project participants often obtained dismissals of charges, so, too, did those not in the project. And while the study found statistically significant differences in dismissals or their equivalent for the two groups (72 percent for the diverted group versus 46 percent for the control group), it warned that many convictions in nondiverted cases were for petty violations, not for crimes. When they compared rates of actual convictions for crimes and jail sentences between the two groups, Baker and Saad reported no significant differences.[28] Thus, while they reported some differences in the ways diverted cases were handled and some benefits to participants, they also found that participation did not increase the likelihood that an arrestee could avoid the stigma of a criminal record.

Interviewing defendants and their attorneys, Baker and Saad found that a substantial minority was not interested in pretrial diversion because they felt they could get as good a deal in court without accepting the probationlike conditions imposed by the program. The majority, however, believed that without the program "not many would be ACD'd" (adjourned in contemplation of dismissal), a judgment that is not supported by the study's findings.[29]

Baker and Saad examined arrest records for a one-year follow-up period and found no significant differences in rates of rearrest between the experimental and control groups. (Thirty percent of the experimentals were arrested in comparison with 33 percent of the controls.[30]) Even when they compared only those in the experimental group who had been designated successful with all those in the control group who had obtained dismissals without CEP, they found no meaningful differences. The study also explored the frequency of rearrest but here, too, found no evidence of program benefit.

The study identified a number of reasons why, despite the good intentions and dedicated work of its staff, CEP had such limited impact. These reasons are similar to those found in other jurisdictions; most notably, those in the diversion programs would have been treated leniently had they not opted for diversion.

Project Intercept—San Jose, California

In 1971, before the results of its two pilot programs in New York and Washington, D.C., were in, the Department of Labor announced a plan to support diversion programs in nine additional cities.* These "second-round" projects were also labeled experiments, and the Department of Labor earmarked several hundred thousand dollars for their evaluation.

One site selected was San Jose, California, a sprawling metropolis of 450,000. The sponsoring agency for Project Intercept, as the diversion program was called, was the Foundation for Research and Community Development, an independent, nonprofit agency that administered on a contractual basis a variety of social service programs in the city. This group con-

---

*The locations were: Santa Rosa, San Jose, and Hayward, California; Atlanta, Georgia; Baltimore, Maryland; Boston, Massachusetts; Minneapolis, Minnesota; Cleveland, Ohio; and San Antonio, Texas.

sulted local judges, public defenders, and prosecutors, who together hammered out the program's philosophy, eligibility requirements, and organizational structure. Because it was sponsored by the Department of Labor, Project Intercept focused on unemployed and underemployed arrestees. At the insistence of local judges and prosecutors, participants were also limited to those charged with lesser misdemeanors, to those under twenty-eight, and to those with no prior criminal records.

Project Intercept received the first of its three annual grants of $130,000 in early 1971 and began operations immediately. Recruiters were stationed in the country's four municipal courts to meet with potential participants and explain the program. With a recruiter's help, applicants were usually able to obtain a ten-day continuance, during which time the staff could explain the project to them in greater detail. If the staff found an arrestee interested and suitable, they notified the court, the district attorney, and the defense attorney. The district attorney and judges came to trust the judgment of the project's staff—or the staff came to anticipate the judgments of the DA and judges—and usually accepted their recommendations. Once accepted, an arrestee began three months of "career development" counseling, consisting of remedial education, referral to job training programs, job placement, and personal advice.

From 1971 to 1974, San Jose's Project Intercept handled an average of 200 cases each year. Participants were usually between the ages of seventeen and twenty-one, slightly more likely to be female than male, often black or Chicano, and usually unemployed or unenrolled in school. Virtually all of them were charged with petty theft or shoplifting.[31]

If during this ninety-day period a participant "showed progress" (usually measured in terms of keeping appointments) and was not rearrested, the project would recommend that charges be dropped. During its early years, the San Jose project recommended dismissals for roughly 90 percent of its clients, although the court followed its recommendations only one-quarter of the time. A project report commenting on this high refusal rate noted that judges in San Jose were "opposed to formal submission of any recommendation [by the diversion program]," viewing it as something akin to an "improper influence" on the discretionary authority of the court.[32]

On the other hand, judges did not disapprove of diversion per se; they apparently regarded Project Intercept as a type of "preconviction probation," which allowed the court to obtain guilty pleas and which at the same time might benefit defendants by reducing the severity of their sentences. As a consequence, the project staff tended to define "success" in terms of keeping participants out of jail after conviction. Over the years, judicial insistence on conviction increased, and in 1978, after seven years of operation, the director of Project Intercept noted that only a handful of participants ultimately obtained outright dismissals. Thus, this diversion project did not do what presumably diversion is all about, provide arrestees with an alternative that allows them to avoid criminal records. And, while the project could claim that few of its clients were eventually sentenced to jail, this was due mainly to the fact that it selected only those charged with petty offenses and possessing no record.

As Labor Department funding neared its three-year limit in 1972, Project Intercept began casting about for other sources of funds. From the beginning, its director, Richard Boss, had linked his program with other community action programs, and as a result he had developed a broad base of support. He was able to marshal an impressive list of backers, and in 1972 the County Board of Supervisors agreed to continue the program. Between 1973 and 1976, the county provided financial support at the level established by the Department of Labor, until its share of local revenue-sharing funds was slashed. Just two months before it was to cease operations permanently, Boss located still another source of funding, the Comprehensive Employment and Training Act (CETA). In fact, because of his program's reputation, Boss was soon able to convince the CETA board to expand his operations. Since 1976, Project Intercept has depended on CETA funding to maintain its operations, and during the Reagan administration its fate has generally followed CETA's.

At the time of my research, Boss was satisfied with this arrangement, at least as long as CETA funds were available. By remaining an independent, nonprofit corporation supplying services on a contractual basis, Project Intercept avoided the complexities that permanent governmental status would bring—the civil service mentality. The contract arrangement provides for flexibility in working conditions and staffing policies. The project was able to keep odd hours and hire a young, streetwise staff. Although the staff was not as well paid as civil service employees in San Jose and had no job security, Boss contrasted their enthusiasm and dedication with what he viewed as the stuffiness of local probation officers.

Boss even viewed institutional insecurity as a blessing in disguise. Counseling, he maintained, is draining and exhausting; after a year or two, counselors burn out and are ready to depart. The low salaries and lack of job security are a benefit: they encourage young staffers to move on to other jobs after gaining experience and confidence.

The shift in funding from the Department of Labor to county taxes caused no significant change in the orientation and operations of Project Intercept. However, the shift to CETA did lead to major alterations. The initial goal of reducing rearrests among participants was all but forgotten by the project. Boss explained this metamorphosis as follows:

> Now we don't measure [success] in terms of dismissals, recidivism and the like, but whether our clients get jobs or obtain a GED [General Education Degree]. From a CETA point of view, it's irrelevant whether someone gets a dismissal or is rearrested. As long as they have a job, it's OK. We don't even bother to keep records of these other things now.[33]

This shift took Project Intercept even further away from pretrial diversion than this quote indicates. Roughly one-half of its clients were obtained from sources other than the arraignment roster in the courts, such as schools and youth centers, and only tradition and concern with young first offenders kept the program involved with pretrial arrestees at all. If funds are cut back or other sources of clientele prove to be more fruitful, Boss acknowledged, the connection with the courts could be cut off altogether. Despite its continued public image as a pretrial diversion program, Project Intercept has all but abandoned diversion. Although it was never very successful at helping clients avoid incarceration or conviction—the original reasons for diversion programs—it ceased even to make much of an effort to do so. Despite all this, its longtime director argued that the project served a useful social and criminal justice function: it helped people obtain jobs. With the demise of CETA under the Reagan administration, even this goal was lost.

Although those court-referred participants in Project Intercept apparently believed that diversion offered them an alternative to a substantial fine or jail term, this benefit was more apparent than real. The district attorney and the courts allowed the project to enroll only those arrestees charged with petty offenses, who would otherwise have received only a small fine or a suspended sentence, and only occasionally agreed to drop charges. In contrast, the project's three-month probationlike supervision was not necessarily desirable. The director acknowledged that his program did not really keep anyone from jail, but nevertheless justified it on the grounds that fear of a possible jail term was an effective "hammer held over the heads" of his clients. It is, he explained, "a powerful incentive to get people to straighten up."

Even though this may be an effective rehabilitation technique—and early on the project stopped maintaining files that could provide answers—such a practice did not necessarily protect the interests of the accused. Clients were recruited prior to their arraignment, which meant that if interested, they, in effect, waived the opportunity to confer first with an attorney. Since the recruiter portrayed the program as a beneficial alternative to conviction and a long jail term, an unwary defendant might have been tempted to agree to acknowledge guilt and participate, and, as a result, he would lose the opportunity to explore possible defenses or challenges to the arrest. Focusing as it did on people charged with petty offenses, the program could easily latch on to people whose charges might be dropped outright if they had not opted for what was offered to them as a benefit.

The New Haven Pretrial Diversion Program

Between 1972 and 1978, a pretrial diversion program operated in New Haven, Connecticut. Designed for youthful first offenders charged with minor (misdemeanor) offenses, the program offered counseling and job placement to a selected number who participated at the discretion of the prosecutor. Participation automatically suspended prosecution for ninety days, at the end of which, if conditions established for the program had been met, the prosecutor dropped charges.

This program was sponsored by the New Haven Pretrial Services Council, a group composed of representatives of the criminal justice system, which was established in 1972 for the purpose of upgrading the pretrial process. The council's founding director was Mark Berger, a recent Yale Law School graduate who had previously been the legal adviser to New Haven's police department, where he had developed a reputation as a thoughtful and innovative adviser.

Like most other early diversion programs, the program in New Haven was based on an informal understanding among local criminal justice officials, an arrangement that presented few problems since the officials were represented on the council and most of them knew Mr. Berger personally. Also, like many other diversion programs, New Haven's program was financed by federal funds, in this case LEAA action grant funds. The New Haven Foundation, a private charitable group, supplied a matching grant. The state's LEAA grant provided $70,000 annually for three years, and during this period the program was expected to obtain an increasing share of its funding from the city.

The New Haven program identified a long list of goals in its grant application: to use "low intervention social service techniques to help stabilize a defendant's employment status in the hope that this will help stabilize his social behavior, to reduce the caseload burden. . . ; provide effective rehabilitation services to offenders immediately after their apprehension; and reduce recidivism among affected offenders and channel as many as possible into productive areas."[34]

In addition to the executive director, the staff consisted of a social worker who served as a project coordinator, a "career developer," and two ex-offenders who served as counselors. Students from area colleges also functioned as part-time counselors.

The staff offered a two-step program: an initial stage of counseling, followed by a career development stage. Counseling consisted of sympathetic listening and encouragement, ad hoc advice on how to cope with problems of daily living, and occasional referral to other social service agencies. If participants were found to be receptive to this enterprise—usually measured in terms of punctuality and appearance at scheduled appointments—they were then assigned to the career developer, whose task was to help them locate a job or reenroll in school.

The program obtained its clients by having one of its counselors review police reports on the new arrestees just prior to arraignments in court. The counselor would then approach potential clients in the lockup facility and briefly explain the nature of the diversion program to them. If an arrestee expressed interest in diversion, the counselor would then ask the prosecutor to allow him into the program and agree to a three-month continuance. If a prosecutor agreed, judges would almost always follow suit.

Despite twenty to twenty-five new arrestees appearing in court each day, half of whom met the program's initial eligibility requirements, the program averaged fewer than one new client per day. There are several reasons for this. Initial interviews were hurried affairs: one counselor had to talk with a number of potential clients and confer with prosecutors, all within a few minutes. Both the defendant and the prosecutor had to make on-the-spot decisions about participation and, when in doubt, tended to say no. In some cases, prosecutors refused to agree to diversion, although the arrestee met all formal eligibility requirements. Many eligible arrestees were never contacted by the program because they had been released on recognizance immediately after arrest and had waived arraignment, the stage at which the program staff contacted prospective clients.

However, the major problem was similar to the one experienced by New York's CEP. Most arrestees knew that they would be released without bail and that there was a good chance the prosecutor would drop charges outright. At worst, they reasoned, they would have to plead guilty and receive a suspended sentence or a small fine. A ninety-day program involving supervision and counseling and no guarantee of dismissal was no great bargain.

During the second year of the project, it was evaluated by a team of Yale Law School researchers, who criticized the hasty and incomplete interviews. In response, the project sought to attract more clients by improving its interviews and broadening its counseling offerings, but it did not appreciably increase its clientele. During the first two years, clients averaged twelve per counselor, well below the target of twenty-five.[35] This extremely expensive ratio was never significantly altered, despite the efforts of the project's energetic new director, Daniel Ryan.

The Yale study also found that the program had only marginal success in achieving any of its goals. Its few participants did not significantly reduce court congestion and, in fact, may have added to delay by requiring special treatment. The Yale team found that none of those in a retrospectively constructed control group received a jail sentence of any length, and that the only person in the study to receive a jail sentence was one of the participants in the diversion program. Typically, a sentence for those in both groups was a suspended jail term or a fine of twenty-five dollars or less. Furthermore, the apparent success of the program in getting charges dropped was probably due to the fact that the project sought out good risks and the prosecutor agreed to divert only weak cases. Finally, the report showed, the per client cost of the program was five times more than the estimated average cost of postconviction supervision through probation.[36] Still, the results of the assessment were encouraging enough to convince the state to provide additional LEAA funds and to convince the prosecutor to liberalize his eligibility requirements in an effort to allow the project to attract more clients.

But the program still could not attract enough participants, and, unable to convince the prosecutor to admit people charged with felonies, the staff continued to concentrate on a handful of people who, for the most part, would have been treated quite leniently by the court without their intervention. By 1976, the Yale Law School researchers and New Haven Foundation officials had become discouraged with the program, and staff positions began to be filled through patronage. And when in 1977 the city was unwilling to pick up even a small share of the funding, the state refused to supply any more LEAA funds. In 1978, the project quietly went out of business.

During its six years of existence, the New Haven pretrial diversion program gained something of a national reputation as a model diversion program. Despite its favorable publicity, based in large part on the fact that its directors were extremely bright, dedicated, and articulate spokespeople for diversion, the record recounted above is rather bleak. The program failed to attract many clients. Indeed, when it finally ended its operations, a number of local defense attorneys and judges were still unaware that it had existed.

Assessing the Pretrial Diversion Movement

However impressive pretrial diversion programs are in theory, they accomplish very little in practice. At best they reassess the value of their activity and move out of the diversion business, as in New York and San Jose. At worst, as in New Haven, they become expensive encumbrances on the courts and finally wither away.

Why, despite serious efforts and substantial funds, has diversion had so little success? One fallacy of the movement was that it proceeded to plan new enterprises as if the primary concern of the criminal justice system were to conform to diversion's goals. Defendants were supposed to welcome diversion as a benign alternative to an alienating courtroom. Prosecutors were expected to divert less serious cases so they could spend more time on more serious cases. Judges were expected to divert people because diversion led to rehabilitation. It apparently did not occur to the reformers that such things might not happen.

Presented as a therapeutic device, diversion was easily adapted as something of the opposite. It became part of a continuing game, to be exploited by players pursuing their own goals. Even the programs themselves got caught up in this process, vying for clients, funds, and credibility, and shifting as needed to succeed.

One of the most glaring shortcomings of diversion was its failure to involve defense attorneys. Diversion quickly came to be shaped by prosecutors, who dominated access to programs. Proponents of diversion did not fully appreciate the conflict inherent in the criminal justice process. Rules, procedures, continuances, and delays are inevitably turned to instrumental end, and diversion was no exception. It became one more weapon to be used by prosecutors. It sought to impose an administrative organization on the courts, one that is incompatible with their fundamental nature and organization.

This incompatibility is evident when the short history of diversion is seen in light of the much longer but equally disappointing history of juvenile courts. Seventy-five years ago, juvenile courts were hailed as a great reform, created in order to separate juveniles from adults and divert them into a more humane system. Yet the experience of juvenile court has not met this expectation.[37] Enthusiasts of diversion have failed to consider the possibility that it would follow a similar pattern and that, in so doing, it would produce new problems akin to those of juvenile courts.

Conclusions

New York's CEP, San Jose's Project Intercept, and New Haven's diversion program failed at the important stages of initiation, implementation, evaluation, and routinization. Each of these several failures will be examined.

Initiation. Emerging as it did during a period of increasing crime, pretrial diversion quickly came to be seen as a way of meeting the problems of overcrowding in the courts and prisons. It became the solution to a crisis even before the nature of that crisis was properly diagnosed and long before it was carefully evaluated.

In part, this was due to factors unrelated to the merits of diversion itself. In the early 1970s, diversion became LEAA's most heavily promoted court-related innovation. LEAA spent its discretionary funds on diversion, required states to spend theirs, underwrote the cost of numerous conferences, and subsidized numerous studies. Within the space of a few short months, an interesting but untested idea was transformed into a magic solution to all court problems. In promoting pretrial diversion, LEAA and other backers were anything but the careful, sober experimenters they purported to be.

Implementation. Judges and prosecutors did not actively oppose diversion since it neither cost them anything nor required them to cede any authority. Defense attorneys could continue to counsel clients; prosecutors retained final authority as to who could participate; judges were not required to automatically dismiss charges against participants. Once diversion programs were established, court officials were not required to do anything differently *if they did not want to*. Indeed, many prosecutors came to regard diversion as an alternative penalty for marginal offenders.

Diversion programs often found themselves short of clients and sought out weak cases to bolster their enrollments and to justify their high expenditures. "Skimming off the cream" came to be a widely known term referring to the practice of selecting weak cases likely to lead to dismissals and hence to program successes. Reports on both the New York and New Haven diversion programs revealed that very few if any of the diverted arrestees would have gone to jail (and many would have had the charges dropped outright or at least not have been convicted) had they taken the standard route through court. In their quest for clients and success, diversion programs focused on those who needed them least.

Evaluation. Nowhere are the failures of the diversion movement so glaring as in its evaluation. This is doubly ironic since diversion was most heavily promoted during a period when LEAA was responding to criticism of wasteful practices, and diversion was the most thoroughly "evaluated" of all its programs. There is literally a bookcase full of evaluations of diversion projects, and there are even several book-length studies evaluating the evaluations. During the 1970s, Abt Associates of Boston, the National Center for State Courts, the American Bar Association, and several regional research organizations all received substantial financial support from LEAA, the National Science Foundation, and other sources to review the evaluations of pretrial diversion programs.[38]

Despite their focus on the better programs—including those in New York, San Jose, and New Haven—the reviews found that none of the evaluations was properly conceived and executed, and both the Abt and ABA studies held that no firm conclusions could be drawn about the effects of any diversion program. The Abt report indicated that the question, "Is pretrial intervention meeting its goal of reducing recidivism?" could not be answered because eight of the nine programs studied had not followed through on their commitments to collect data on control groups.[39] Roberta Rovener-Pieczenik reached similarly disheartening conclusions in her report for the ABA.[40]

Ten years after diversion was first promoted as a bold new experiment, and after millions of LEAA and Department of Labor dollars had been spent on hundreds of experimental projects and a great many studies, LEAA finally commissioned a careful evaluation. The Court Employment Project in New York City was the subject, and the research was directed by Sally Baker of the Vera Institute. This study, discussed earlier in the chapter, confirmed many of the worst suspicions of a growing number of skeptics.[41]

Routinization. Pretrial diversion grew from a handful of pilot programs in the late 1960s into a full-fledged movement in the 1970s. But by the end of the decade, its bright hopes had nearly burned themselves out. Many programs experienced the fate of New Haven's: quiet death, once federal funds were exhausted. Others, as in San Jose, have survived a bit longer in name only because their directors have been especially adept at locating new sources of money. A few, as in New York, have undertaken candid self-assessments and consciously sought to improve their operations. Very few diversion programs have become permanently institutionalized with steady and secure funding. Many of these surviving diversion programs have purposefully sought to isolate themselves from the courts. As the director of the San Jose program expressed it: "We are not part of the system; we're an alternative to it." While such an attitude no doubt enhances a staff's esprit de corps and appeals to some of its clients, it is purchased at a high price. Isolated outside programs are especially vulnerable once federal funds or special grants are exhausted and local sources of support are required. Not being a part of the criminal justice network, such programs are often unable to mobilize enough local support to sustain them beyond their initial stages and thus are left to die or shift direction, especially as competition for shrinking funds increases.

Duplication. This chapter has linked program failure to the problems of overselling, lack of understanding by outsiders, and insecure funding. It also illustrates an important point made in the introduction: courts are not what they appear to be. For instance, supporters argued that diversion provided an informal alternative to encumbrances of formal adjudication. Yet, in fact, the courts are not highly formal institutions. Diversion programs failed in part because many courts had already adopted flexible and informal alternatives on their own.

Contrary to myth, prosecutors are often willing to drop charges when the interests of justice are compelling. In any given U.S. city, roughly one-half of all misdemeanor arrests and a sizable number of felony charges are disposed of without conviction. And among those convicted, a large majority avoids incarceration. In many courts, domestic disputes are routinely referred by prosecutors to social workers. In petty cases, defense attorneys are often able to obtain dismissals or reductions in charges by getting their clients to make restitution, see a psychiatrist, enroll in school, or join the army.

In New York, a number of seemingly serious charges were reduced to misdemeanors or dropped altogether because the court learned that the parties involved were acquaintances and had reconciled, or that stolen goods had been returned.[42] The Vera evaluation of CEP suggested that the alternative to diversion and dismissal was outright dismissal or, at worst, conviction on a noncriminal charge (a violation). The alternatives generated by the court itself are, in fact, the most important reasons why diversion programs have failed. The programs often duplicate in time-consuming and expensive fashion what the courts have long been doing informally and at little additional cost.

A Modest Contribution

In many cases, formal diversion programs have been preempted by the more effective practices of informal diversion and de facto and de jure decriminalization. Yet the diversion movement may have accelerated and influenced this informal change. Like the pretrial release movement, diversion's greatest impact may be indirect. Although diversion programs have rarely succeeded, the movement has helped to legitimize the use of alternatives to criminal penalties. While prosecutors have routinely dropped petty charges against the affluent or those whose pastors or ward leaders come forward with a good word, now they may act more systematically and equitably. Diversion was originally conceived as one part of a broad concern with overreliance on the criminal sanction. To the extent that the diversion movement has helped to promote this concern, it may have made a valuable contribution. Police now routinely ignore public inebriation and possession of small amounts of marijuana, and use of alcohol detoxification centers has expanded.

Pretrial Diversion

I offer these concluding observations as further evidence of the convoluted nature of the courts and the problems of effective *planned* change within them. Even as their proximate objectives were not reached, the very existence of diversion programs may have precipitated significant by-products. But such realization is not meant as consolation; it does not exempt us from the hard task of finding out how to initiate and implement *planned* change.

Neighborhood Justice Centers

What pretrial diversion was to court reform in the 1970s, neighborhood justice or dispute settlement centers are becoming in the 1980s. They are the new cure-all. Such diverse supporters as Chief Justice Warren Burger, Attorney General Griffin Bell, Senator Edward Kennedy, and the American Bar Association have promoted them. These centers permit laymen trained in mediation and informal therapeutic techniques to resolve petty disputes—criminal and civil—involving family and neighbors, consumers and merchants.

Supporters of these centers argue that they will allow the courts to devote more time to serious cases; that the high price of attorneys and long delays have made courts inaccessible to all but the affluent; and that the formalism of the courts works against mediation, conciliation, and compromise. A 1978 Department of Justice publication enumerated the goals and benefits of neighborhood justice centers:

· To establish in the community an efficient mechanism for the resolution of minor criminal and civil disputes that stresses mediation and conciliation between the parties in contrast to the findings of fault or guilt which characterizes the traditional adjudication process.

- To reduce court caseload by redirecting cases that are not appropriate for the adversarial process.
- To enable the parties involved in the disputes to arrive at fair and lasting solutions.
- To serve as a source of information and referral for disputes that would be more appropriately handled by other community services or government agencies.[43]

These goals are much more ambitious than the aims of pretrial diversion. The neighborhood justice center movement envisions a substantial alteration in the structure of the courts and the creation of an entirely new institution. Who could be against increased fairness, easier access, decreased costs, lasting settlements, and preventive actions? But are the goals realistic? Are there hidden drawbacks? Can this attractive theory be put into practice?

Observers have suggested that case overload is an intrinsic feature of courts, particularly in civil matters, and that delay may even be functional in that it encourages out-of-court settlements.[44] They question whether informality and nonbinding solutions are efficacious in the absence of a tight-knit and culturally homogeneous community.[45] Others assert that courts rarely *resolve* conflict, that at best they only *process* it. If someone is not satisfied in one forum, he or she is likely to pursue the matter in another forum.[46] The creation of an additional layer of legal institutions may result in an increase in conflict and cost, not in the expected decrease.

Finally, it is interesting to note the similarity between the proposed informal dispute resolution centers and the old-fashioned justice of the peace, an office now extinct in many places. There is irony in eliminating an informal legal institution and having it reappear in slightly different form.

Pretrial Diversion

There were good reasons for questioning the justice of the peace system, and there are, I submit, equally good reasons for skepticism over the new proposals. They have the potential for some of the same kinds of abuses found in many JP courts.

Some skeptics maintain that the new informal alternatives often fail to appreciate the realities of the lower trial courts, picturing them as formal and rigid institutions. Yet, as we have seen, lower courts are on the whole remarkably flexible and adaptable when considering so-called petty disturbances—the very types to be sent to the new centers. Neighborhood justice centers, like diversion programs a decade before them, may do little more than duplicate in an expensive and cumbersome way what is already provided by these courts. And if our concern is that court officials are remote and unresponsive, there is no reason to believe that an institutionalized alternative would fare any better. Experience with pretrial diversion and the less than spectacular history of the American justice of the peace suggests that this most recent romance with informality may be no more successful than its predecessors. I raise this not to prejudge—certainly the legal process is in need of experimentation and improvement—but to urge caution and the need for a skeptical and historical perspective. I am not, however, optimistic that such advice will be followed.

In 1978, Attorney General Griffin Bell opened three federally funded "experimental" neighborhood justice centers in Atlanta, Kansas City, and Los Angeles with a great deal of fanfare, and LEAA quickly made vast sums of money available to replicate them in other locations. Congress jumped on the bandwagon, passing legislation authorizing their creation. The ABA continues to ballyhoo the idea. And, increasingly, private foundations are supporting them.

As with pretrial diversion, all this activity is taking place in the absence of any firm knowledge about actual effectiveness,

111

with little careful inquiry into the handful of historical anteced-ents and with little skepticism. Again we see the emergence of a "movement" fueled by high ideals and the burning desire to find solutions even before the problems are properly under-stood.

Despite the "experimental" label given to the federal three-city effort as well as other numerous programs, and de-spite the fact that the Justice Department commissioned an expensive evaluation, federal officials have not worked to assure careful, experimental evaluations.[47] A proposal to treat these experimental programs as true experiments and randomly as-sign would-be clients to neighborhood centers or leave them to their own devices was explicitly and firmly rejected by the Department of Justice.

At a ceremony opening one of these centers, Attorney Gen-eral Griffin Bell did not simply indicate that the department was undertaking an interesting but as yet unproven idea. In-stead, he unveiled a "bold new idea" that was going to solve many problems of the courts. The first steps into the new area were not tentative and provisional, but overstated. They were accompanied by moral fervor, certainty, a blitz of publicity, and exaggerated promises. I am afraid that by 1990 the "de-cade of neighborhood justice centers" will have ended with the same yawn that ended pretrial diversion in 1980. This would be unfortunate, for even if the solutions are misguided, the problems they address are real and they deserve better.

Despite this inauspicious beginning, there is cause for opti-mism. In 1981 the Ford Foundation established the National Institute for Dispute Resolution (NIDR) to develop and assess a variety of alternative forms for resolving disputes. The Insti-tute is headed by Madeleine Crohn, former director of the Pre-trial Services Resource Center. Its mandate includes explora-tion of alternative ways to handle criminal and civil cases.

Pretrial Diversion

If the Institute can overcome the twin problems that plague foundations and government agencies alike—the tendency to exaggerate and prematurely promote ideas, and the resistance to criticism and careful evaluation—it is in a good position to make a difference. Its director is well respected in the larger policy-making community, yet is also knowledgeable about the mundane details that can make the difference between success and failure in particular programs. If its aims for affecting the criminal courts are modest and incremental, the Institute can be an effective agent of change.

To this end it might encourage reflective commentary on how much due process is due those with various types of grievances; stimulate assessment of practices and alternatives in operation in Europe to determine whether they have applicability to the North American setting; reconsider a role for the much-maligned lay magistry in the United States; support research that maps the existing informal institutions for handling disputes already in place in the United States; review alternatives already well institutionalized in a limited setting (for example, commercial arbitration) to determine whether they have potential for broader applicability; and sponsor research to determine if use of alternatives actually results in cost savings that are passed on to clients by their attorneys.

Chapter 4

SENTENCE REFORM

The Origins of the Modern Sentence Reform Movement

Cesare Beccaria argued against flexible sentencing practices in his famous treatise, *On Crimes and Punishment*, published in England in 1767,[1] and Jeremy Bentham quickly joined his crusade.[2] They showed that while 200 crimes were punishable by death, there were in fact relatively few executions, and they criticized an indiscriminate process of discretionary punishment that they claimed neither deterred crime nor fostered respect for authority.[3] They proposed in place of that system one of moderate but fixed penalties that would be dispensed evenhandedly. Though Beccaria and Bentham had many adherents in England, most notably the legal reformer Sir Samuel Romilly and Quakers repelled by the death penalty, their philosophy failed to achieve any momentum until the nineteenth century, when it was embraced by business interests.[4] The rulers of England remained committed to the *in terrorem* effect of capital punishment, coupled with frequent last-minute demonstrations of mercy.

But during the mid-nineteenth century, the philosophy of

fixed punishment based on the seriousness of the crime gradually took root in England and the United States. The penitentiary was an essential ingredient in this new scheme, replacing the two most common penalties, capital punishment and transportation to prison colonies. By the late 1800s, this was the dominant sentencing approach in the two countries.

But even as this philosophy gained formal acceptance, realities of practice served to restrict its application. Limited resources and the need to exercise control over prison inmates led to the modification and uneven application of this new philosophy. Governors were prevailed upon to pardon large numbers of prisoners because of overcrowding. As early as 1808, criminologist David Fogel points out, "the Newgate Penitentiary in New York was granting so many pardons as to make discharges equal to commitments, while Ohio simply pardoned convicts whenever the population rose above 120 in number."[5] The promise of early release for "good time" in prison became a tool for maintaining inmate discipline. New York passed the first good-time law in 1817, and by 1869 good-time provisions, allowing for as much as a 25 percent reduction in sentence, had been enacted in most states.

Despite these modifications, dissatisfaction with fixed sentences continued to grow. At the turn of the century, a new theory emerged in the United States, this time rooted in the new disciplines of psychology and sociology.[6] This theory regarded the criminal as a social deviant in need of rehabilitation rather than punishment. Since it held that length of imprisonment was best determined by experts who were capable of diagnosing each prisoner's needs and responsiveness to treatment, this argument was used in support of legislation that ceded much of the judge's sentencing discretion to prison authorities and parole boards.

New York once again led the way; in 1876 it became the

first state to adopt an indeterminate sentencing scheme based on rehabilitative principles. The New York State legislation did not give complete sentencing authority to prison administrators and parole boards; judges were still required to set minimum and maximum sentences. But within these limits, prison officials had latitude for discretion. Flexible or indeterminate sentencing—fitting sentences to the offenders rather than the offense—became one of the programs of the Progressive movement and subsequently was widely embraced. By 1922, thirty-seven of the forty-eight states had adopted some form of indeterminate sentencing, and forty-four had created parole boards.[7] The most complete practical expression of this rehabilitative principle emerged from the creation of the California Adult Authority in the 1940s. The Authority was granted broad power to set prison terms, and its express statutory purpose was to tailor sentences to fit the offender's requirements for "treatment." Thus, in a one-hundred-year period, the theory of sentencing had zig-zagged, from discretionary to fixed to discretionary practices, suggesting a dialectic without synthesis.[8]

But by the 1960s, the modern theory of indeterminate sentencing based upon the rehabilitative ideal had come under attack. There was growing consensus that a theory advocating treatment and rehabilitation was untenable in the absence of proven diagnoses and cures for criminal conduct. Some opposed indeterminate sentencing because it violated the requirement that those similarly situated be treated equally. Others charged that indeterminate sentencing fostered haphazard and lenient sentencing, which undercut the deterrent function of the criminal law. Still others charged that it permitted parole boards to mete out excessively harsh sentences to whomever they wished.

There are significant differences among contemporary advocates of sentencing reform. Proponents of mandatory minimum sentencing seek to limit judicial discretion in order to obtain a minimum punishment, while advocates of determinate sentencing seek to restrict judicial discretion in order to foster certainty and consistency in punishment. The underlying theory of the former emphasizes deterrence, while the underlying theory of the latter rests upon a concern with parity and commensurability. Both theories permit some leeway in application, but both stand in sharp contrast to those views they are designed to supersede. These rest on rehabilitative assumptions and justify a wide range of minimum and maximum sentences within which a parole board is free to act when specifying actual terms.

In this study, I am interested in the similarities of these two new and distinct sentence reforms. Both oppose many existing flexible sentencing provisions and practices; both seriously question the rehabilitative ideal; both seek to restrict judicial discretion; and, of greatest interest for purposes of this study, both attempt to alter long-standing sentencing practices.

Sentence reformers appeal to the logic of their theories. Here, I propose to consider these theories in action. This functional examination reveals what a formal analysis of policy principles cannot. Despite possibilities to the contrary, judicial and parole board discretion has not been the culprit in fostering leniency and disparity that it is commonly thought. And in spite of obvious appeals, current reforms often miss their mark by unwittingly fostering other and less visible forms of disparity, even as they pursue policies to the contrary. Further, reflection on the limits of the language of law reveals that however hard they try, legislators who prescribe minimum or fixed sanctions to particular offenses run the risk of attaching a precision to language that simply does not exist. No matter how carefully defined, an offense collapses a wide variety of behavior into a single category and in so doing masks differences likely to be important. Discretion recognizes this simple fact.

This chapter examines the impact on the courts of four recent sentence reforms: the Rockefeller drug law in New York, mandating stiff minimum sentences for drug possession and sale; the Bartley-Fox gun law in Massachusetts, requiring a one-year mandatory minimum prison sentence for possession of a deadly weapon; the Michigan mandatory minimum sentences for narcotics sale and illegal possession of firearms; and the massive restructuring of California's penal code, transforming its long-established indeterminate sentencing system into a determinate system. These case studies reveal a tendency for reformers to exaggerate the expectations of their proposals and make questionable contrasts between the complex reality of current practices and the simplifications of their formal theory. The chapter concludes with a proposal for a pragmatic approach to sentence reform that would approach specific problems more directly than comprehensive reforms and permit continued judicial discretion albeit under the watchful eye of appellate courts.

The Rockefeller Drug Law

Nelson Rockefeller was not engaging in hyperbole when he called his approach "brutal." As originally introduced in January 1973, his proposed amendments to the state penal code specified a mandatory life sentence for anyone convicted of the sale or possession of narcotics, prohibited plea bargaining, and eliminated parole. The governor also proposed mandatory life sentences for those convicted of committing violent acts while under the influence of narcotics and a 100 percent property tax on drug dealers, so that upon conviction, the state could seize all of an individual's private property. Informants were eligible for a $1,000 bounty if their information resulted in convictions for narcotic violations. The bill's only concession to leniency was that youthful offenders between the ages of sixteen and eighteen could be paroled after fifteen years.

New York's chapter of the American Civil Liberties Union called the bill "a frightening leap towards the imposition of a total police state."[9] Robert McKay, dean of the New York University School of Law, president of the Legal Aid Society, and past chairman of the Commission on the Violence at Attica, commented: "It's completely counter to everything the civilized world has been working for."[10] But their reactions were predictable. What was remarkable was the adverse reaction of hard-liners on crime. One well-known "tough" judge predicted that if enacted, the proposals would cause the courts to stop functioning.[11] Another was concerned that "we'd have to multiply by many, many times the number of judges, nonjudicial personnel, and ancillary personnel." The New York State District Attorney's Association, the New York City police commissioner, the Association of the Bar of the City of New York, the New York State Bar Association, the New York Judicial Conference, the New York State Probation Officers Association, and the five district attorneys of the City of New York all went on record in opposition to the bill. People from all segments of the criminal justice community testified in opposition to the governor's proposals. Their opposition focused on two aspects of the bill: its failure to differentiate between large and small drug dealers, and the heavy burden it would place on the criminal justice system.

The great unknown was the effect these new measures would have on police practices, particularly in New York City, where drug traffic was heaviest. Shortly before the governor made his proposals, the New York City police department had adopted a policy of concentrating on big dealers rather than on the thousands of petty user-dealers who haunted the streets. This experience showed that little was accomplished by periodic sweeps of the streets for users. If anything, mass arrests tied up officers on trivial cases and limited their pursuit of the smuggling and distribution bosses. Rockefeller's proposals would force them to take petty cases seriously. The proposals reduced distinctions between the occasional user and the leader of an international drug ring; if convicted, both could receive identical sentences.

The courts voiced strong opposition. In 1972, only 8 percent of all narcotics convictions resulted from jury trials; the majority of convictions was obtained by plea bargaining. Most of these plea-bargained cases involved petty users. In contrast, major dealers rarely obtained concessions from prosecutors, and if convicted, they usually received stiff sentences. Therefore, officials familiar with narcotics cases saw no advantage in Rockefeller's proposals. Furthermore, they pointed out, if prosecutors could not plea bargain in the large number of petty cases, the courts would soon be swamped. Without the incentive of a chance for sentence leniency, virtually everyone charged with possession or sale would insist on a trial, and each trial would take an average of six or seven days, rather than the four hours for a plea bargain. Since the bulk of the cases would be in New York City, the courts there would feel the greatest strain. One estimate was that the trial rate would jump from 9 to 85 percent of all narcotics cases, which led Mayor John Lindsay to present his own politically motivated estimates: he argued that the number of judges would have to be increased from 79 to 370 simply to keep abreast of the anticipated increase in trials. The mayor's Criminal Justice Coordinating Council pointed out that implementation of the new law would cost $400 million for construction of courtrooms and prisons alone, and another $150 million to $200 million annually to staff and operate them.[12]

Governor Rockefeller responded that these petty "practical" objections could easily be overcome. If more judges and court space were needed, he was prepared to obtain them. He requested a $55 million appropriation to expand the state's criminal court system and to cover the cost of one hundred new judges.

The promise of additional funds and courthouse positions mollified those who saw a patronage bonanza. The governor openly acknowledged the tactical importance of patronage. When asked whether the promise of so many new supreme court judgeships was a "carrot," he replied: "No, this is not a carrot. This is a tool. This is the mechanics [sic] to meet the needs."[13] The governor recommended that the legislature create one hundred new civil court judges who would be selected (conveniently) by gubernatorial appointment rather than by election and then "temporarily" transferred to the new narcotics courts. Rockefeller's response to critics was: "It's only unconstitutional if you call them a Supreme Court judge. Call them something else, it's constitutional."[14]

The New York Times editorialized that Rockefeller's proposal was a "politically attuned harangue that threatens to make a bad situation worse."[15] During legislative debate, the governor's political ambitions were frequently criticized, and one legislator warned his colleagues that they were "going to louse up the criminal justice system in the name of getting this Governor elected next year."[16] Rockefeller was known to harbor presidential ambitions, and he knew he needed to gain support among the Republican Party's right wing. (Actually, the governor did not run for reelection in November 1974, having been named vice-president by Gerald Ford after the collapse of the Nixon presidency.)

The 1973 drug bill was not the first attempt by the governor to exploit the drug issue for political purposes. During his 1966 gubernatorial campaign, Rockefeller had promised compulsory hospitalization programs for addicts convicted of crimes. The Narcotics Addiction Control Commission, established with great fanfare in 1967, absorbed vast amounts of money —perhaps as much as $1 billion—to finance construction of secure treatment facilities. The scheme was a costly failure; addicts were usually released uncured after a brief period of incarceration. Some observers argued that this new proposal was intended to divert attention from Rockefeller's earlier debacle.[17]

The governor's scenario of a drug plague did not conform to reality. Statistics on narcotics use in New York City at the time did not reveal any massive upsurge; to the contrary, there had been a definite decline. By 1973, narcotic deaths and serum hepatitis cases in New York City had dropped markedly from a peak in the first part of 1971,[18] and admissions to methadone clinics had leveled off.[19] Social workers, police officers, and others familiar with the drug scene also had the impression that heroin use was declining prior to Rockefeller's amendments.[20]

Stiff drug penalties already existed under New York State law, but they were not invoked. Before the passage of the new law, a person convicted of the sale or possession of sixteen ounces of narcotics could be sentenced to life imprisonment, with the possibility of parole only after serving fifteen years. Sale or possession of lesser amounts could result in sentences of up to twenty-five years. Mandatory minimum sentences had been in effect since 1951 and had been raised in 1966. Nevertheless, the Joint State Legislative Commission on Crime, in a 1967 study, revealed that only 2.5 percent of convicted drug sellers received the maximum penalty. The mandatory minimums had been ignored or bypassed by judges and prosecutors.[21] Thus, to achieve substantial increases in incarceration, all one had to do was press prosecutors not to plea bargain and judges to give the sentences already provided for—and in some cases mandated—by the law. This failure to use existing legislation, more than anything else, suggests that the Rockefeller effort was essentially symbolic, rather than a serious effort to tackle a real problem.

Provisions of the Act. The near unanimous opposition of defense attorneys, judges, prosecutors, and police was all but crushed by the Rockefeller juggernaut. On May 9, 1973, only four months after outlining his new plan, Rockefeller signed the "nation's toughest drug law." The law created three categories of class A drug offenses: A-III, A-II, and A-I felonies, which provided for mandatory minimum penalties of one, six, and fifteen years, respectively. The maximum sentence in all categories was life imprisonment, and all parole releases required lifetime supervision. The act differed from the initial proposal in only one important aspect—it permitted some plea bargaining. But there was a draconian twist. Individuals charged with the more serious A-I and A-II felonies could plead guilty to a lesser A-III felony and thereby escape the mandatory six- and fifteen-year sentences; but a person charged with an A-III felony could not bargain at all. Thus, the law permitted the most serious cases to be treated no more harshly than the least serious cases.

Once opposition in the legislature was overcome, Rockefeller turned to implementing the act, aware that some of his most vociferous critics would be administering it. In his address at its signing, Rockefeller condemned these critics for being part of a "strange alliance of vested establishment interests, political opportunists and misguided softliners who joined forces and tried unsuccessfully to stop the 1973 drug law."[22] He warned: "We are creating the strongest possible tools to protect our law abiding citizens from drug pushers—providing that the police, the district attorneys and the courts throughout the state are willing to use these laws vigorously and effectively."[23] In short, he was saying that these critics would bear the blame if the law failed.

Effects of the Act. Rockefeller's concerns were not without foundation. Said one well-placed police official shortly after the signing: "We could destroy Rocky's program in two months, just by making a lot of arrests—enough to paralyze the courts. There are people who would like us to do that, to give the legislators what they asked for."[24] The threatened subversion never materialized. After adoption of the new law, the New York City Police Department continued its policy of concentrating on large-scale drug dealers and eschewing enforcement of the law against petty user-dealers.[25]

The judiciary lost no time in expressing its distaste for the new law. Obliged to sentence a defendant to a mandatory sentence, one judge stated:

> This legislation, since its enactment, has instead of alleviating the ills of drug abuse and addiction, created chaos and much consternation among those who are responsible for the administration of criminal justice. The administration of criminal justice has been debilitated because of the apparent reluctance of law enforcement and judicial authorities to implement the unduly harsh provisions contained in these laws.[26]

But there is no evidence that judges sought to sabotage the act. Prior to its implementation, only 5 to 10 percent of all convicted drug offenders received sentences of more than one year, while after adoption, one-third received minimum sentences of one year or more, and 80 percent of those convicted of class A drug felonies in 1975 were sentenced to one year or more.[27] (Those few who avoided prison sentences did so because they were classified as informants or came under the law's youthful offender exceptions for sixteen- to eighteen-year-olds.)

Court administrators had predicted that it would take five years to adjust to the new law,[28] and that the system might suffer permanent paralysis. While paralysis did not prove to be a correct diagnosis, thrombosis did appear to set in. Case backlogs rose dramatically and only began to level off after two years, after some of the harsher provisions of the law were repealed[29] and after a doubling of criminal court parts (i.e., judge, stenographer, bailiff, etc.) in New York City.[30]

The costs for this expansion were substantial: more than $40 million annually for salaries alone and additional millions for new capital expenditures. The estimated costs to the courts for processing the approximately one thousand "new" prison sentences produced by the law between 1974 and 1976 was roughly $40,000 per offense, a figure that does not include the $8,000 average cost for one year in prison.[31]

Many opponents predicted that prosecutors would be the officials most likely to nullify the act, pointing out that the ban on plea bargaining and the mandatory sentences could be circumvented by preindictment charge bargaining.[32] This did not happen. While prosecutors were hindered by a rising backlog of cases, this was due to a lack of experience in the new narcotics courts. Figures showing no changes in police arrest practices and substantial increases in sentences suggest that neither police nor judges nor prosecutors subverted the new law.*

Despite this apparent good-faith administration, the law accomplished few of its major aims. In fact, fewer defendants

*Prior to the drug law's enactment, the New York City police had stopped targeting small drug dealers. This policy was maintained after the law's passage. While the act provided sharp penalties for even petty drug involvement, full police enforcement would have completely overloaded the system. See Drug Law Evaluation Project, "Crime Committed by Narcotics Users in Manhattan, Staff Memorandum" (New York: Association of the Bar of the City of New York, May 1976).

charged with drug violations were sentenced to prison in the three years following passage of the law than in the three years preceding it, and fewer defendants were convicted in the years after passage than before.[33] This seemingly anomalous situation arose in part because of the long delays caused by the new law, and in part because the new law made it more difficult to obtain a conviction.[34] The class A indictments were the main cause of this backlog. While constituting only 75 percent of the new law's cases, they represented 90 percent of the pending new case backlog in New York City.[35]

In response to increased motions, the New York City judicial dismissal rate in drug cases rose from 6.8 percent in 1972 to 21.3 percent in 1975. This suggests that after passage of the Rockefeller law, judges became more sensitive to allegations of defects in arrests and indictments than they had been in the past. This inference is strengthened by the marked decrease in dismissal rates in early 1976, immediately following the repeal of the harshest provisions in the Rockefeller drug law.[36]

Between 1973 and 1975, the trial rate for drug cases in New York City jumped from 6.5 to 15 percent.[37] While these figures are significantly lower than the estimates, it must be remembered that as passed the law did allow for limited plea bargaining, something that Rockefeller's original proposal had prohibited. The habitual-offender section of the 1973 law also increased the burden on the courts; the nondrug trial rate in New York City increased from 6.6 percent to 10.1 percent between 1973 and 1975,[38] most of these increases owing to the act's habitual-offender provision.[39]

The new law created a number of reasons why defendants would invoke the adversary process and prefer to risk the uncertainties of trial rather than accept the more predictable outcome of a guilty plea. Judges did not sentence those convicted after pleas any more harshly than those who were convicted after trial.[40]

While the law clearly increased the likelihood of prison terms for those convicted of selling large amounts of heroin, it was still most lenient toward those at whom it was most clearly aimed, A-I and A-II offenders. The Rockefeller law permitted those charged with the most serious offenses to get the best deals. Sentencing data reveal this pattern: a major violator originally indicted on an A-I and A-II felony who pled guilty to an A-III felony was just as likely to receive the minimum one-year term as a petty offender originally indicted on and convicted of an A-III felony.[41] Offenders who possessed or sold less than one-eighth of an ounce of heroin were as likely to serve the same sentence as those who had been found with a pound.[42]

This glaring inequity, coupled with the rising backlog of cases and the decline in conviction rates, eventually led to the move to repeal the plea-bargaining restriction on A-III cases.[43] In 1976, as a way of coping with this backlog, the special narcotics prosecutor in New York City announced that he was going to ignore the clear intent of the law and allow defendants charged with A-III felonies to plead guilty to class A misdemeanors and thus avoid indictment and the mandatory minimum one-year sentence. With Nelson Rockefeller safely out of the way as vice-president, the legislature amended the 1973 drug law to permit class A-III felony defendants to plead guilty to lesser offenses. This move had an immediate and dramatic effect. In New York City, the trial rate for felony drug cases dropped from 17 to 9 percent, and for those charged with A-III felonies, the rate plummeted from 34 to 6 percent.[44]

Several lawsuits unsuccessfully challenged the constitutionality of the Rockefeller law.[45] One of them was filed on behalf of a thirty-eight-year-old woman who had no prior criminal record and had been convicted for possession of one ounce of cocaine and sentenced to life in prison.[46] This suit and others like it forced legislative reconsideration of the act, and in 1979 the state legislature repealed the major provisions of the act and ordered a complete review and readjustment of sentences handed down under it.

To its credit, the legislature acknowledged that the critics had been correct; the law had had little measurable deterrent effect on narcotics use, it was astronomically expensive to administer, and it led to harsh sentences for marginal offenders and little increase in punishment for major offenders. The opponents had been wrong in only one respect: all the evidence points to a good-faith enforcement effort by police and judges. Only when they found the act to be overly burdensome did prosecutors move to thwart it, and even then they did so in an open and aboveboard manner.

Nelson Rockefeller's last present to the state was an embarrassment that was quickly undone once he had removed himself from the scene. Still, he left a costly legacy. New York State finds itself with a much enlarged and increasingly expensive judiciary. Their raison d'être now removed, these new judges have been assigned to handle other types of cases, and while no doubt they are put to good use, they are a permanent and costly reminder of Rockefeller's folly.

The Bartley-Fox Gun Law

Boston has long enjoyed a reputation as a relatively "safe" city. Thus it was with some surprise that in 1974 the *Uniform Crime Reports* revealed that the rate of reported robberies in Boston was second only to that of Detroit among the nation's large cities. David Bartley, speaker of the Massachusetts House, joined with J. John Fox, a recently retired judge of the probate court and one-time criminal court judge, to propose a sentencing bill that would impose a mandatory minimum of one-year imprisonment, without possibility of suspension, parole, or furlough, on anyone convicted of carrying and possessing a firearm without a license and gun permit. Both men possessed a talent for publicity, and their proposal quickly gained widespread support.

Their bill had two distinct aims: to punish offenders more harshly and to circumscribe the sentencing discretion of judges. Both stemmed from what they regarded as judicial cowardice. Judge Fox was not reluctant to criticize his brethren on the bench: "Who was it who said that war cannot be left to the generals? Crime in this nation can no longer be left in the hands of the judges. We can no longer afford the discretion we have placed in them. We can no longer allow it in the matter of crime."[47] As eventually adopted, Bartley-Fox—as the law came to be known—provided for a one-year minimum sentence in selected gun cases but did not prohibit prosecutors from reducing a "carrying" charge to a simple charge of "possession," a crime not covered by the minimum sentence.

Despite the bill's potential for disrupting the courts by encouraging defendants to seek trials or to engage in protracted plea bargaining, there was no notable opposition to Bartley-Fox from elements of the criminal justice system. Resistance primarily stemmed from the legislature's black caucus and supporters of stronger gun control.[48] Only the hesitancy of the governor, Francis Sargent, prevented the bill from being adopted in record time.*

*Interested in more stringent gun control, Governor Sargent refused to sign the bill and returned it to the legislature, requesting that amendments be added to ban Saturday night specials—cheap handguns. Sargent's initiative was eventually defeated, and the measure was enacted and put into operation on January 1, 1975. Some

Bartley-Fox placed no restriction on gun ownership but did require that all owners obtain a license or firearms identification card (FID). On the surface, it simply provided a penalty for failure to obtain the necessary papers, an action of insignificant value in itself. But Judge Fox thought otherwise. "There are," he said, "the several hundreds of thousands of youngsters who have been committing crimes because we have taught them that crime does pay. . . . We have misled them because there has not been sentencing for crime."[49] The legislature found his argument convincing and adopted his proposals more or less as he presented them, although as enacted the law did not apply to "youngsters" (that is, juveniles).[50]

Effects of the Law. Since Bartley-Fox first went into effect, it has been closely watched, and it is possible to make some tentative assessments of its effects. As in New York, critics predicted that officials would find ways to circumvent the law and pursue business as usual. They predicted nonarrest by police, charge adjustments by prosecutors, and dismissals and charge reductions by judges. Although these dire predictions did not materialize, there is evidence that the criminal justice system has adjusted its behavior to absorb much of the law's sting. Bartley-Fox had some deterrent effect immediately following its adoption but then appeared to decline, and its long-range deterrent effect is a matter of question. Several factors account for its decreased efficacy: fewer people are actually being charged with the offense carrying the mandatory sentence; judges and juries are increasingly reluctant to convict on this charge; and, probably most important, the new law is redundant in serious armed robbery cases carrying sentences in excess of one year. Still, Bartley-Fox has not met with outright

attributed the law's legislative success to the fact that the powerful Massachusetts gun lobby (Springfield is the home of the Winchester Arms Manufacturing Co.) hoped that passage of the act would prevent a stronger gun-control law from passing.

official subversion and clearly has produced some expected benefits.

A study of the first year of Bartley-Fox concluded that it had effected a small but demonstrable reduction of violent crime, and that the court system had meted out the one-year sentences relentlessly.[51] A follow-up study of the law after three years found a noticeable drop in gun robbery rates but an increase in nongun armed robberies. In the third year of Bartley-Fox, there was an upsurge in the use of guns in robberies, suggesting that the law's effect may have been temporary—a response to the initial publicity.[52] Several other measures reinforce this interpretation. After passage of the law, there was a reduction in gun-induced wounds as well as a reduced homicide rate, while at the same time there was a marked increase in assaults with knives, stones, and clubs.[53] Still, on balance, Bartley-Fox has had a positive impact on crime.[54] Assaults have not diminished but have shifted to less lethal forms.

Critics of Bartley-Fox had predicted two responses to it. Either the law would be circumvented by police and prosecutors or it would be systematically enforced and the courts would become clogged. Neither came to pass. The agents of the criminal justice system enforced the law in a flexible manner, and the courts' operations continued almost unimpeded. An explanation of why Massachusetts averted the problems that befell New York in the wake of the Rockefeller drug laws and Michigan after adoption of similar gun legislation lies in the special nature of the Massachusetts court system, the few arrests, the absence of organizational resistance to the law, and the very limited nature of the law.

Massachusetts has a two-tier trial court system. The lower or district courts have exclusive jurisdiction over minor offenses. They share jurisdiction with the trial or superior courts over all misdemeanors and those felonies carrying penalties no greater than five years. District courts also have arraignment authority for all cases. As a consequence, they hold preliminary hearings in serious felonies, and at times defendants charged with felonies use this opportunity to plead guilty to lesser offenses and remain in the district court. Defendants convicted in a district court have the right to a trial de novo in superior court, where they can request a jury rather than a bench trial.

Bartley-Fox had a significant effect on the ways gun cases were disposed of in these two courts. In 1974, immediately prior to its passage, only 18 percent of all gun-carrying cases in the Boston district courts were appealed or bound over to superior court; in 1975, immediately afterward, this rate jumped to 47 percent and has remained high.[55]

In district courts, most of the trials take no more than half an hour, and many require only ten minutes. The process is informal; there is no stenographer; police prosecute most cases. The district court traditionally removes the incentive for appeal by offering reduced charges, considering extralegal factors not easily considered in more formal proceedings, and at times offering lenient alternatives to conviction and incarceration. These courts hold frequent trials, have no backlog to speak of, and, with some few notable exceptions, are well regarded by defense attorneys.

With adoption of Bartley-Fox, the ability of the district courts to offer this rough substantive justice in gun-carrying cases has been severely restricted but not altogether eliminated. Prior to Bartley-Fox in 1974, 16 percent of all gun-carrying cases were resolved by means of a "continued for dismissal" (an eventual dismissal if the defendant steered clear of the law for six months),[56] while after adoption of the law, this alternative was eliminated. Also prior to Bartley-Fox, district courts in Boston would often dismiss a case but nevertheless "punish" the defendant by assessing court costs; this practice, too, was ended by the new law.[57] District court judges began sentencing gun-carrying offenders to minimum one-year prison terms, and this practice in turn led defendants to appeal to the superior courts in record numbers for trials de novo.

While the proportion of cases appealed has risen dramatically, the absolute increase has been small. At most, Bartley-Fox has increased the superior court case load in Boston by no more than seventy cases annually. Several factors account for this. Police now make many fewer arrests on gun-carrying charges, and judges in district courts increasingly dismiss or acquit defendants on these charges.[58] In Boston, the primary target of Bartley-Fox, the number of arrests for gun carrying declined significantly after passage of the law. In 1974, there were 218 arrests; in 1975, 186; and in 1976, 168.[59]

There is little evidence to suggest that the initial decline was due to conscious police nullification.[60] Police officers were generally confused about the scope and specific provisions of the new law, often arresting on possession instead of for carrying, but they were not deliberately filing the lesser charge in order to avoid the mandatory sentence.[61] When police discovered weapons outdoors where there was no ambiguity about the law, they invariably made an arrest and charged a carrying violation.[62] When in doubt, the police initially indicated a preference to undercharge.[63] Since the Massachusetts courts eventually ruled that Bartley-Fox did not apply in one's home or place of business, the police cannot be faulted for their charging decisions.[64]

An examination of the courts where the police control the charging decision indicates that sympathetic defendants are increasingly being charged with the less serious possession offense.[65] In the Boston municipal courts where the judge decides the charge based upon an officer's arrest report, this pattern was not found.[66] But even when judges determine the charge, there are regional disparities. In the western part of the state, judges admitted substituting the less serious possession charges for carrying charges.[67]

As in the New York courts under the Rockefeller law, judges reviewing challenges under Bartley-Fox also tend to be more appreciative of the intricacies of due process and the standards of proof than they once were.[68] But the success rate of defendants in the district courts has not appreciably changed from what it was prior to the law. Previously, successful defendants had their cases continued without a finding. The new law prohibited this practice, and judges were forced to rule on the legal merits of the case.[69] Thus, one by-product of Bartley-Fox is a lower rate of conviction owing in some cases to sympathy and in others to a heightened concern for due process.[70] Ironically, one of the major consequences of this antioffender law has been a prodefendant response.[71] Still, those convicted on the charge now do time, whereas before adoption of the law they usually did not.

These changes have not occurred solely because judges are now more solicitous of the rights of the accused. In light of the harsh penalties, defense attorneys are much more willing to challenge the charges in district courts, insist on a trial and, if they lose, to repeat all these efforts in superior court. Occasionally, a compelling case for leniency can be made. Interviews with Boston area defense attorneys, prosecutors, and judges revealed that "prosecutors do take into account . . . the fact that the defendant's record did not warrant one year in jail."[72]

Because of all this, the predicted avalanche of new work has not materialized. In Boston in 1974, there were eleven gun charge trials; in 1975, eighteen; and in 1976, seventeen.[73]

134

Sentence Reform

Most contested cases are filtered out early by dismissals, reductions of charge, and plea bargaining. Despite the mandatory sentence, if the case involves other separate charges, where the defendant faces punishment in excess of the one-year minimum, there remains considerable incentive to bargain. This is especially true in cases of armed robbery, where judges traditionally hand down sentences longer than one year. But overall, plea bargaining has dropped from the pre-Bartley levels. In fact, many defense attorneys and prosecutors stated that they engaged in no or little plea bargaining.[74] If a gun case goes to trial, the probability of an acquittal is substantial. Between 1974 and 1976, the percentage of acquittals in bench trials in gun cases increased from zero to sixty, and in jury trials from fourteen to seventy.[75]

Reflections on Bartley-Fox. On balance, Bartley-Fox has produced effects consistent with the expectations of its proponents and has yielded up few if any of the major problems predicted by its opponents. Available early evidence suggests that it has had a mild deterrent effect, although it is impossible to separate out the influence of the increased sanctions of the law from other factors. The reductions in cases involving armed violence that followed the law may be due in part to the temporary blitz of publicity that accompanied the new law or to the highly publicized reorganization of the Boston Police Department, which took place concurrently with the introduction of the new law. Still, it appears certain that the law has had some deterrent effect.

Our primary concern in this study is with the impact of the new law upon the courts. Here, too, while there have been noticeable adjustments, the law has not precipitated the outright adaptation and nullification predicted by its opponents and found elsewhere. In Boston, the courts have been able to adjust to the new law quite easily, in large part no doubt because so few cases are involved. In western Massachusetts, there appears to have been greater adaptation by police, prosecutors, and judges. While there have been a variety of adaptations that have reduced the impact of the new law on the courts, there has been no widespread and systematic effort to nullify the law. While Bartley-Fox has increased the number of defendants charged with carrying a firearm, only a small proportion of them are convicted, and the number of gun charge arrestees who actually go to prison remains quite small. As of April 1978, after four full years of operation, there were only fifty-eight persons serving prison sentences for violation of the Bartley-Fox gun law,[76] and in only five of these fifty-eight cases was violation of the new law the only offense for which the offenders were serving time. In many of these cases, the gun charge was an add-on to a more serious charge. Furthermore, sentences for more serious felony convictions that included a firearms conviction as well were not more severe and if anything may have been shortened.[77] If the intent of Bartley-Fox was to sentence anyone convicted of carrying a gun to a minimum period of imprisonment, it is achieving its purpose; but if the intent is to lengthen sentences for felons who use firearms in the commission of crimes, or to increase the likelihood of conviction for use of guns, it has had little impact.

The Michigan Mandatory Minimum Sentences

In the 1950s, the Michigan legislature enacted a mandatory twenty-year minimum sentence for the sale of narcotics. In fact, few offenders have received this "mandatory" sentence. Researching the impact of this new law, Donald J. Newman found that judges and prosecutors felt such a sentence was too harsh for all but a handful of professional dealers. Newman reported that in the typical case involving a small user-dealer, prosecutors and judges willfully circumvent the law by reducing charges. "It is clear from the universality of the practice," Newman concluded, "that the purpose is not to individualize the consequences of conviction but to nullify the legislative mandate on the grounds that it is inappropriate in the usual case."[78]

A similar law—the Felony Firearms Statute—imposing a mandatory two-year minimum prison sentence for possession of a firearm while engaging in a felony, was enacted by the Michigan legislature in 1976.[79] It provides for sentences of five and ten years for second and third offenders. Although the law does not prohibit plea bargaining, many prosecutors, including the district attorney of Wayne County (greater Detroit), announced at the time of enactment that they would not plea bargain in these cases.

The bold announcement of a new era of gun law enforcement notwithstanding, actual practices in the court remain the same. Before the law went into effect in 1976, 56 percent of those convicted of armed robberies received sentences in excess of two years; in the six-month period after passage (1977), the rate was 55 percent.[80] The typical sentence for assault convictions both before and after implementation of the law was no time in prison. In felonious assault, the anticipated direction of sentences was reversed: in 1976, 86 percent of those convicted received no prison sentence, while in 1977 the corresponding figure was 82 percent.[81] In only one type of case is the law consistent with expectations. Between 1976 and 1977, there was a noticeable increase in sentences over five years in length, suggesting that the provision for recidivists, or the tough stance of the new district attorney, was having some effect.

In the face of both a harsh new law and an announced new policy of no plea bargaining, how was it that imprisonment rates did not increase as expected? The answer is that judges have routinely subverted the new law by dismissing gun charges, substituting lesser offenses, refusing to convict on the gun charge, and, in cases involving multiple charges, reducing the usual minimum sentence on the primary offense to compensate for the two additional years mandated for gun charges. Plea bargains are now often arranged by direct conference between defense counsel and judge. Outright nullification of the law is rare, but an increased appreciation for requirements of due process, which leads to dismissals and charge reductions, is not.

These reactions by judges arose and persist out of long-standing expectations among court officials as to what constitutes reasonable sentences. Judges generally believe that the sentences meted out prior to the passage of the new law were reasonable and justify their discretion and flexibility as responses to what they see as an overly rigid sentencing scheme.

Ironically, the county prosecutor's accompanying policy to restrict plea bargaining had the opposite effect. After adoption of the new law, the scope of plea bargaining changed.[82] The refusal of prosecutors to engage in charge bargaining in firearms cases has encouraged judges to engage in sentence bargaining. Thus, the new law has driven underground—or at least into chambers—that which once was an aboveboard practice. One judge commented: "Now everybody is involved in a kind of conspiracy to undermine [the law] without getting caught."[83]

Sentence Reform

The California Determinate Sentencing Law

During the 1950s and 60s, California was frequently hailed as a national leader in the treatment and rehabilitation of prisoners. One source of this view was the state's indeterminate felony sentencing law, which required county courts to impose sentences that commonly provided for high maximum penalties (for instance, fifteen years for a single charge of daytime burglary), but also allowed a centralized agency (known as the "Adult Authority" after 1944) later to fix the actual terms. These were usually only a fraction of the statutory maximum. In granting parole the Authority emphasized its commitment to rehabilitative considerations and, at the same time, prison officials publicized their efforts to develop therapeutic, educational, and work programs for prisoners.

The winds shifted during the 1970s, but again California is being hailed as a national leader. Its Uniform Determinate Sentencing Law of 1976 proclaimed "the purpose of imprisonment is punishment,"[84] not treatment and rehabilitation. The new law abolished both the Adult Authority and indefinite sentences. Now most prisoners have definite sentences imposed by judges who must select from a narrow range of options prior to imprisonment. The new law appears to adopt a philosophy at odds with that reputedly animating the old law. It appears to reflect concern that penalties be proportionate to the gravity of the offender's conduct, in contrast with the old law which seemed to emphasize rehabilitation and its underside, incapacitation.

At first glance California seems as deeply committed to its new determinate sentencing system as it was a few years earlier to its indeterminate system. But first glances can be deceptive. Legislative majorities are often comprised of coalitions with quite different aims, and the laws they pass may fail fully to guide the decisions of those who must implement them. Such is the case with the new California law.

While shifts in intellectual currents and the frustrations of practical experience often give rise to replacement of one set of views with another, the fact is that officials must continue to cope with a great variety of concerns. Regardless of which views gain temporary salience, the others are not altogether forgotten. Practices have a way of separating themselves from their animating theories, and theories have a way of encroaching upon diverse practices such that correspondence between theory and practice is often difficult if not impossible to perceive. For instance, California's original parole and indeterminate sentence laws were promoted by those who sought more statewide uniformity in sentencing and not by committed advocates of the rehabilitative ideal. And during the height of the rehabilitative ideal, the Adult Authority was arguably as much concerned with equalization of disparate sentences as it was with tailoring individualized sentences.

Under the old law, even minor infractions were punishable by long prison terms, and actual terms were in theory based upon a judgment about rehabilitation while in custody. In practice, however, only a handful of offenders actually served their maximum terms or had their terms determined as a result of treatment and rehabilitation. In 1965, for example, the median time actually served by prisoners for offenses carrying a maximum of life imprisonment ranged from a high of 5.4 years (second-degree murder) to a low of 3.0 years (burglary).[85] And in spite of its mandate to assess responses to treatment, the Adult Authority usually followed a well-established practice that translated the law's indeterminate sentences into a more or less fixed fraction of the maximum sentence allowable. There were exceptions, of course, notably with sympathetic and political prisoners, but for the majority of cases, sentences were, in a statistical—although not subjective—sense, fairly predictable.

In the 1970s, people united in their opposition to the Authority often had different criticisms. Some challenged the rehabilitative ideal underlying indeterminate sentencing, while others—pointing to the occasional brutally long sentences—opposed the Authority for its caprice.[86] Criticism was also directed at the Authority's unwillingness to set a discharge date until just before it was prepared to grant parole.[87] Despite the Authority's predictability, the absence of enforceable and public guidelines for setting parole and release created intense insecurity among state prisoners, who would do anything to try to convince officials they were rehabilitated. Still, they could never be assured of early release, and had no recourse to effective review if they thought they had been unfairly treated.

Despite this absence of due process, the Authority did at times probe into disparities caused by plea bargaining and make an effort to equalize sentences by setting terms on the basis of the facts of the case rather than the formal offense category.[88] Still it could do little about local practices that led to considerable variation in the rate at which offenders were committed to prison, perhaps the greatest source of sentence disparity across the state.

According to some, the Authority was "seen as a valve which opens to relieve pressure."[89] Here the Authority was "cajoled, manipulated, begged by the Department of Corrections . . . to have enough prisoners but not too many" and it closely followed monthly prison population projections in order to adjust parole rates accordingly.[90] However, another observer of the Authority remarked that it was a "sticky valve at best" since the Authority was not under the control of the Department of Corrections and followed its own course of action.[91]

The benefits of these functions should not be exaggerated. The Authority's two-member teams heard up to twenty-five cases a day.[92] Prisoners were not entitled to counsel, did not have an opportunity to rebut inaccurate information, could not cross-examine witnesses, and were unable to present witnesses on their own behalf. Typically, officials reached their decisions after little more than a moment's whispered conference; rarely did they state reasons.[93]

These cursory practices suggest that the Authority's decisions were largely routine responses to a few factors. Still, the Authority could not withstand the charges of caprice, nor could the formal (although rarely invoked) philosophy of rehabilitation escape the philosophical shifts. Opponents of the Authority grew in numbers throughout the 1960s, and in the 70s, after the California Supreme Court began to review its decisions, mounted a full-scale attack.[94]

The Move Toward Determinate Sentencing. By the mid-1970s, support for the Authority had evaporated. Liberal supporters replaced the rehabilitative ideal with theories justifying retribution, deterrence, or isolation.[95] Conservatives felt the Authority was too lenient. Prosecutors came to oppose it, for fear that its actions would invite still more appellate review of sentences. Even those who supported the Authority wanted major administrative changes.

In a series of cases during the early 1970s, the California Supreme Court ordered the Authority to rationalize its practices and set penalties in proportion to the offender's criminal record.[96] These decisions encouraged Senator John Nejedly, chairman of the Senate Select Committee of Penal Institutions, to seek a review of the state's sentencing laws. An aide, Michael Salerno, set about the task and found widespread support for major revisions based upon a determinate sentencing system. The notable exceptions were the judiciary, the Department of Corrections, and the Adult Authority.

This opposition was formidable, but the courts continued to hammer away at the Authority.[97] The 1976 decision, *In re Stanley,* found wanting the Authority's most recent administrative reforms, and is generally regarded as the catalyst that precipitated the legislative initiative that led to the new law.[98]

The New Law. The law defined four categories of prison terms for all felonies (excepting capital crimes) and three tiers of punishment within each category: sixteen months, two years, three years; two, three, and four years; three, four, and five years; and five, six, and seven years.[99] "When a judgment of imprisonment is entered," the law reads, "the court shall order the middle of the three possible prison terms, unless there are circumstances in mitigation or aggravation of the crime."[100] Sentences can also be increased well beyond this range, but only if the prosecutor pleads and proves such factors as "enhancements" (for example, carrying or use of a weapon or inflicting serious bodily injury).[101] The law also permits consecutive sentences when multiple charges are proven.

The Impact of the New Law on the Courts. Although the new law has drastically shortened maximum sentences in single-charge cases,[102] early evidence suggests that actual practice under the old and new systems are not so very different. This is because the presumptive sentences are based on the average sentences actually served under the old system and because of the tendency of courts to maintain their old work routines. The major difference is that terms are now known at the outset.

The new law did nothing to curb the discretionary authority of prosecutors and may even encourage plea bargaining. This possibility led to early predictions that the law's principal effect would be to enhance discretionary practices much less amenable to control than were the judgments of the Adult Authority. While there was considerable incentive to bargain in prison-bound cases under the old law (charges established minimums and maximums which had to be respected), the Authority could and at times did look beyond formal charges when fixing actual terms. In contrast, the new law explicitly ties sentence to charge and other factors which are easily controlled by the prosecutor.

As yet, however, a major expansion of plea bargaining has not occurred, although one study found that prosecutors in Alameda County now engage in specific term bargaining, something that was not possible earlier.[103] Another report by Casper, Brereton, and Neal found considerable variation in the frequency with which enhancements were argued and accepted.[104] They also found that some prosecutors are vigilant in their pursuit of secondary charges while others concentrate on only the most serious charges or are willing to negotiate over them. These authors conclude that current practices mirror local patterns established before adoption of the new law, although enhancements and similar devices do appear to provide prosecutors with a new set of resources for negotiation. Overall, Casper and his colleagues conclude, "despite its expressed concern for increasing uniformity, [the new law] has not by any means produced a narrow range of relatively equal sentences."[105] To some extent this is a matter of judgment since the new law permits considerable discretion in sentencing.

In their examination of the new law, Sheldon Messinger and his associates at Berkeley found that median terms decreased in the year or two following adoption of the new law, but that there was a marked increase in the rate of commitments to prison (as opposed to being placed on probation).[106] They also found considerable regional variation in rates of imprisonment. For instance, one study in this project reports that after the new law "there was at least as much variation in the use of imprisonment, between counties, as there had been before the law came into effect—which is to say, quite a lot."[107] While disparity is an elusive concept, the magnitude and systematic nature of these variations suggest that they stem from differences in the propensities of local officials. This research also suggests that there are no ascertainable *major* changes in the variability of length of prison terms under the new law.[108] The explanation for this lack of change suggested by the study is "that the typical terms eventually served by those committed to prison in California [under the old law] did not fluctuate that much," due in part to the fact that terms were administered centrally . . . by the Adult Authority.[109] Now, sentencing is presumably more structured, but it is also decentralized and under the control of local officials.

During the past two decades California officials have been caught between two cross-cutting pressures: the desire to reduce prison crowding and the desire to respond to calls for harsher penalties. Both pressures have had an impact, at times simultaneously. In the late 1960s, corrections officials prevailed upon Governor Reagan to encourage the Authority to open its safety valve and to parole offenders more quickly. As a result prison population declined. Several years later the Governor called for greater firmness and the prison population increased. However, the recent rises are due more to an increase in the proportion of felony offenders sent to prison than to longer terms. Indeed, for most offenses there has been a decrease in the actual or estimated sentence, a trend that, like that of rising commitments, apparently began before but has continued since adoption of the new law.

This brief review indicates that California's new law has not significantly affected sentences, but that it may have pushed many considerations back to an earlier and less visible stage of the process. At the same time, it has abolished one of the institutions—the Adult Authority—that had the opportunity to take a statewide perspective and had some capacity to equalize sentences stemming from variations in local customs. Thus, many of the old issues remain unaffected by the new law. While the new act created the Community Release Board—later renamed Board of Prison Terms—to review "mistakes" in sentencing, this board has quite limited powers and to date has been reluctant to exercise even those.[110]

Since its adoption there have been amendments to the new law, providing for still longer sentences, although to date they have not translated into widespread increases in actual terms. The new law has also done away with an important moderating practice. Under the old law, in a highly publicized case, a judge could threaten a harsh sentence, knowing that the Adult Authority would set a shorter, appropriate sentence later. The new law strips away this hypocritical device, but it also invites policy to be fashioned in response to unusual cases.

While granting that the new law may be more a product than a cause for a return to a higher rate of imprisonment, it does nothing to discourage tougher sentences. Whatever the causes, now a higher proportion of felons are sent to prison than were sent during the period just preceding the new law. And since adoption, amendments have expanded probation ineligibility conditions and in effect mandated minimum terms in some cases. Per capita imprisonment rates are again on the rise.

Should this trend continue, the state will be faced with some difficult choices. Prison overcrowding and the State Supreme Court's concern with prison conditions may force the state to devise ways to regulate populations. The question is how. Unlike the Adult Authority, the new Board of Prison Terms can do little about crowding. It has only limited authority to review sentences and none to release. Unless its authority is expanded it cannot function as a safety valve. However, there are plans for massive prison construction. Prison authorities are also considering a variety of nontraditional alternative sentences. By expanding the definition of custody to include such programs, they may be replacing one type of discretionary judgment for another. It remains to be seen whether they will be adopted and, if so, whether they will be flexible enough.

This brief examination of California's new law reveals that it, too, allows for considerable discretion and variation. Despite the replacement of one "comprehensive scheme" with another, many old problems remain, and many new problems have been spawned. It is not clear whether the new law has yielded net benefits. When and if the state corrections department expands its range of alternatives and prosecutors' newfound powers over sentences come under close scrutiny, the state may again begin a seemingly circular search for the correct sentencing policy.

Conclusions

The reforms reviewed in this chapter have had three interrelated aims: to strengthen the deterrent effects of sentences, to standardize punishment and, in California—at least according to some—to reduce sentence disparity.

The new laws rarely achieved their aims. Mandatory sentencing plans failed to increase deterrence because they did not increase arrests. Any reduction of a targeted criminal activity immediately after adoption of new laws was short-lived and was probably due more to a temporary blitz of publicity than to the laws' provisions.[111] Moreover, under the old, supposedly lenient laws, serious offenders were already receiving harsh sentences. At times serious offenders even fare better under the new laws. This is simply an unanticipated result of the exercise of discretionary judgment permitted under the new laws.

It is not clear that the reforms have reduced inequality and disparity in sentencing. In California, reduction of sentence variation is more apparent than real. In other states, various practices blunt the seemingly automatic application of mandatory minimum sentences. In Detroit, judges are for the first time actively engaged in sentence bargaining. In Massachusetts, juries are now willing to convict on lesser included charges in order to prevent application of mandated penalties, and judges seize upon evidentiary and procedural technicalities for the same end.

In all the states, the new laws had the effect of increasing sentences for petty offenders rather than for serious offenders, leading to greater, not less, sentence disparity in many cases. Even California's seemingly precise sentence scheme depends upon prosecutors to introduce enhancements and probation ineligibility provisions; prosecutorial propensities to do this vary widely. Furthermore, by shifting responsibilities from the Adult Authority to the trial judge, the prosecutors' power to affect sentencing has been enhanced, and the problem of variations remains.

On the horizon is the likelihood of still greater adaptation. New York has already backtracked. California prisons are bulging and may lead to adaptations creating new forms of discretionary behavior. In Massachusetts, the courts have been able to cope because of the few cases and a general decline in case load, a trend that at any time may be reversed.

It has been reported that the first person arrested under the Massachusetts gun law was a seventy-year-old woman who was afraid of being raped. No doubt this story is fanciful; nevertheless, it illustrates a point too often ignored by reformers. However artfully drawn, laws remain inadequate to capture the complex array of situations to which they will be applied. Fixed or mandated sentences will at times lead to unacceptable results, overly lenient or overly harsh, and officials will find ways to circumvent them. We have seen such responses here.

I do not wish to paint a wholly negative picture of these sentencing reforms. All crimes must be taken seriously. And allegations of judicial discrimination, arbitrariness, and caprice cannot be ignored. However, this chapter suggests that these concerns have not always been taken seriously enough. Proponents of new theories have not adequately addressed the implications of these theories, and too often sponsors have designed legislation more for its symbolic appeal or formal logic than for its actual effect. This is most obvious in New York, where the law Nelson Rockefeller introduced with such fanfare was quietly dismantled by an embarrassed legislature once he left state government.

A Supplementary Explanation of Sentence Reform Policies. This chapter opened with a brief historical review of the cyclical nature of sentence reform. No doubt this oscillation indicates public fickleness, but it also reveals continuing frustration over a serious social problem that yields to no simple solutions and even defies diagnosis. More generally, this constant shifting reveals a deep-seated tension inherent in the very nature of law itself. Law is a fragile instrument, but it is called upon to perform Herculean tasks. It must seek to be universal; yet it is also faced with the need to be particular. It has to be certain and precise; yet on the other hand it recognizes the complexity of human experience and the need to be responsive to particular circumstances. Further, the criminal law is expected to educate, to deter, and to rehabilitate. Each of these diverse needs is reasonable in itself; yet they are not fully compatible.

At best they coexist in uneasy and perpetual tension, moving toward no natural equilibrium. At times, as we have seen, one or another of these concerns gains ascendancy, but the others are also present and seek outlets to assert themselves.

Granting discretion to officials at key junctures of the criminal process is but one of the many devices for coping with these tensions. However, as we have seen, this policy of flexibility breeds a suspicion all its own, and it is not surprising that, periodically, efforts to restrict it gain prominence. But this in turn brings forth a new set of concerns, and the cycle begins anew.

Where insoluble problems exist, we must accept proximate solutions. But such pragmatism is difficult to accept; it appears to signal defeat when the public wants victory, inaction when politicians want bold new initiatives, ambiguity where scholars want clarity. So rather than educating these groups, reformers offer simple solutions, contrasting the promises of formal theory with the complexities and shortcomings of current reality. In the end, such formalism exacerbates rather than ameliorates problems. It is a politics of dramaturgy, which at its worst can widen the gap between what the public wants and what government can deliver.

At the beginning of this book, I argued that the game metaphor was valuable in understanding the operations of the courts. In this chapter, we have seen it operate at two levels. At the level of initiation, change is shaped by the game of electoral politics. At the implementation level, it is shaped by the adaptation and fragmentation of autonomous and often antagonistic agencies. Several sentence reforms examined here also triggered the quintessential courthouse game of plea bargaining. Where they did not, they were short-lived (New York) or have had relatively limited impact (Massachusetts).

The sentence reform movement is not equivalent to the diversion movement. Perhaps because the changes require legislation, states have been slow to move and have been varied in their approaches. All of those changes discussed in this chapter—as well as many others that were not—were and continue to be subject to scrutiny by evaluators, who present their findings to policy makers in timely fashion. No doubt one of the reasons why mandatory minimum sentences have not been more rapidly adopted and widely applied is that careful evaluations of the New York and Massachusetts laws revealed many serious problems. The California law included a detailed monitoring requirement for various state officials and has also precipitated several large-scale independent investigations.[112] Early mixed results, coupled with recognition of the need for long-term research, have led many other states to develop a cautious attitude toward major change. In 1980, Minnesota began implementation of a presumptive sentence act, which creates guidelines that prescribe a zone of sentences for specified offenses but permits judges to deviate from it for good cause. Here, too, the state legislature mandated careful evaluation. Evaluations have played a significant part in the sentence reform movement. They have pointed out limitations, encouraged experimentation, and, most important, acted as brakes on the impulse to adopt comprehensive changes quickly. All this is encouraging. The concern with *impact* of new sentencing plans stands in sharp contrast to the lemminglike enthusiasm associated with the pretrial diversion movement.

The Future Agenda of Sentence Reform. Regardless of the philosophies underlying new policies, the most significant change in recent years has been a move to greater reliance on imprisonment. Incarceration is used much more frequently in the United States than in other industrialized countries, and even within the United States there is significant variation—Florida prisons hold about as many offenders as California's, despite the great population differences.

Central to the debate on sentencing policy should be the extent of reliance on incarceration and consideration of suggested alternatives (fines, work release, community service, perhaps even corporal punishment). But as long as the current debate continues, these alternatives are likely to be neglected and, if considered actively, opposed because they require discretionary judgments and individually prescribed terms.

A pragmatic approach to sentencing should separate it from crime control. I suggest this not because I think there is no connection between the two, but because—with perhaps some few exceptions—within the range of variation we are willing to tolerate, the actions we are able to take are so limited as to be minimal to nonexistent. Although dramatic increases in the length of prison sentences might arguably reduce the incidence of crime, as a society we are not prepared to impose penalties harsh enough to make a difference.[113] Indeed, until quite recently we have been reluctant to impose even marginally increased sentences in white-collar crime, which almost certainly would have some significant deterrent effect.[114]

A reasonable objective of sentencing policy is the creation of a fair and just system of sentencing. But even here there are no easy answers. Indeed, despite widespread concern with sentence disparities, it is surprising just how little careful research and thoughtful inquiry have been directed at this issue. Examination of sentence disparity has tended to focus on the occasional horror story and the potential for abuse permitted under broad laws rather than on careful and systematic inquiry about actual sentences in vast numbers of routine cases. Similarly, discussion of sentence disparity usually begins with the assumption that all variation is undesirable rather than with an exploration of the nature of disparity or what features may be undesirable.

If, for instance, one accepts the position that it is legitimate to consider both the seriousness of the offense and the prior record of the offender in setting sentence, then sentences will always be subject to charges of disparity because there is no natural balance between the two factors. Moreover, we can continue to define and limit the role of the judge at sentencing, but as long as the process of getting a case to the judge remains fluid and flexible, precision and certainty at sentencing will be problematic.

Typically, we are confronted with stories of gross disparities—the forty-year sentence for possession of marijuana in Texas versus the five-dollar fine in Michigan. But, like it or not, many differences must be tolerated under a federal system. Many local courts appear to have developed standard sentences for various types of offenses and offenders, and cases tend to fall within the range prescribed by these informal norms.[115] There appear to be some significant differences across communities; judges in upstate New York may be less tolerant than their counterparts in Manhattan. But such differences reflect different community cultures, which public policy might arguably respect.[116] Also, seemingly precise definitions of offenses can obscure significant variations in seriousness, culpability, and nature of incidents. Much so-called deterioration of charges and disparity in sentences is due to such factors.[117] The law makes no distinction between robbery-as-debt collection by an acquaintance and robbery as confrontation by a stranger; yet police, prosecutors, judges, and victims often do. Nor may it make a distinction between an assault erupting from an alcohol-fueled quarrel between lovers and an assault of a stranger at night; yet here, too, courts often do. While we may disagree with the course of action in any particular case, we must learn the broader lesson here: the law can mask differences when we think distinctions should be made. Ignoring such differences can create disparities of still another kind.

The recent drive to restrict judicial discretion and reduce sentence disparity may have generated a major social problem where none existed before or elevated a limited problem to a major one. If the idea of disparity—as opposed to invidious discrimination—is complex and does not lend itself to clear understanding, let alone simple solution, and if the readily obvious social problems are found in the occasional case, there is a response that is at once less drastic and more likely to be effective.

Appellate review of allegedly unfair sentences would focus on individual cases and procedures and would be particularly well suited to identify and correct clear mistakes. Judges focus on specific problems; judicial decision making is incremental. On a case-by-case basis, courts can move slowly and cautiously, preserving and reinforcing those practices that seem to work and questioning and replacing those that do not. In the process of tackling the problem, they define it. Sentence review is incremental and experimental in a way that comprehensive reform is not and cannot be.

This suggested approach does not eliminate a role for the legislature, nor is it novel. Indeed, this interactive process between courts and legislatures is well under way concerning the question of capital punishment. I believe courts can do much more in this area, but at the same time I acknowledge that the courts have already precipitated dramatic changes. There is no reason to believe that they could not broaden their scope to consider other types of sentences as well. In so doing, they would invite legislatures to consider sentence policy carefully. Legislatures could create standards for sentencing that would be consistent with appellate review of sentencing. Here, the

Sentence Reform

British experience would be a valuable guide, although it too neglects to consider the impact of plea bargaining.[118]

Sentencing is often more complex than expected, and court systems are often more adaptable than imagined. Sentencing problems are better handled by slow and continuous adjustment and experimentation than by comprehensive reforms.[119] For this reason, appellate review of sentences is especially promising, although given the current concerns about our courts, I am not optimistic about this possibility for the near future.

Chapter 5

SPEEDY TRIAL RULES

AND THE

PROBLEMS OF DELAY

ONLY within the past few years has the Sixth Amendment "right to a speedy and public trial" received serious attention from the courts and from reformers. Efforts to enforce this right are of three types: court cases, court rules, and legislation. The first two have been little more than symbolic gestures by the judiciary: at the state level, the third has fared only slightly better. In passing the Speedy Trial Act of 1974, Congress demonstrated a serious resolve to do something about judicial delay. However, to date, formal speedy trial provisions have had limited impact because they have not reached those who move cases through the courts, and because they have done little to alter the incentives of the defense bar which, on the whole, finds delay functional. Still, the recent interest in adopting speedy trial provisions indicates an increased concern over the issues of delay, and there is evidence that this general interest, if not the specific provisions, is having some effect.

This chapter looks first at constitutional interpretation of the right to speedy trial and then moves on to explore the federal court rules dealing with the problem of delay. It examines congressional and state efforts and, finally, studies the problem of delay in detail in an effort to understand why reforms have had such limited effects.

Speedy Trials and the Constitution: The Impact of Supreme Court Decisions

The Sixth Amendment's guarantee of speedy trial has been before the United States Supreme Court only a handful of times. Still, in recent years the Court has spoken eloquently on the importance of this constitutional guarantee, calling it "an important safeguard to prevent undue and oppressive incarceration prior to trial, to minimize anxiety and concern accompanying public accusation and to limit the possibilities that long delay will impair the ability of an accused to defend himself."[1] Despite this ringing affirmation, the Court has rarely found a violation of this right, a record that is nothing short of amazing.[2]

This constitutional double standard is dramatically illustrated in a 1973 case, *Barker* v. *Wingo*, the most important Supreme Court ruling on the issue to date.[3] The Court held that despite a five-year delay in trial, there was no violation of the right to a speedy trial; and that there exists "no constitutional basis for holding that the speedy trial right can be quantified into a specific number of days or months."[4] Instead, the Court developed a balancing test consisting of four factors: length of delay, reason for delay, prejudice to the defendant, and defendant's assertion of his rights. The Court stated that since circumstances varied in each case, a hard and fast rule would be unacceptable. In *Barker*, the court might reasonably have held that the defendant had waived his right to a speedy trial by actually seeking delay. However, by failing to suggest guidelines or a concrete test, the Court in effect washed its hands of the problem. As with most balancing tests, the claims of the individuals gave way to the needs of the state.

Since *Barker,* the Supreme Court has never overturned a conviction for lack of a speedy trial, although some dismissals have occurred in the lower federal courts.[5] The Court's reluctance to act decisively prompted some in Congress to propose legislation, but judges have fiercely resisted such congressional action, at first by informally lobbying against it, then by offering their own court rules as an alternative, and, finally, when Congress did act, by resisting implementation.

Court Rule Making

Traditionally, courts have had considerable freedom to formulate standards for their own internal operations. This license is justified by the theory of the separation of powers and, in the case of the federal courts, is institutionalized by legislation. Under the Court Reorganization Act of 1925, the Supreme Court has the authority to promulgate rules of procedure for the federal courts, which automatically go into effect after ninety days unless Congress intervenes to alter them. These rules are usually adopted only after extended deliberation by judges and legal scholars and after acceptance by the full federal judiciary. Much of what the federal courts adopt under their rule-making powers is also properly within the purview of Congress, although over the years the judges have come to view this authority as something of a right.[6]

In the 1960s, as the President's Crime Commission, the American Bar Association, the Department of Justice, and

Speedy Trial Rules and the Problems of Delay

Congress began actively considering legislation to deal with problems of delay, the federal judges proposed to develop a solution of their own. Working through the Federal Judicial Conference, the Federal Judicial Center, and the Administrative Office of the U.S. Courts, the judiciary mounted its campaign against delay. The judges' solution to the problem was more resources, and during the late 1960s and early 1970s they convinced Congress to appropriate additional funds. Still, hearings before the Senate Committee on the Judiciary in 1971, 1973, and 1974 revealed that the judges were not keeping abreast of their cases. These hearings also found that despite isolated experiments by a few innovative judges, the judiciary was not effectively reducing delay.[7]

Following the 1971 hearings, Senator Sam Ervin and Congressman Peter Rodino introduced similar bills mandating speedy trials in federal criminal cases. The bills required all criminal cases to be brought to trial within sixty days of arrest, unless they came under certain specific exemptions, and provided for dismissal of charges if the court did not meet its deadline.

Although in earlier rulings the courts had in effect left remedies for delay and specifications of fixed time periods to the legislative branch, the federal judiciary fiercely resisted the new legislation. Similarly, both the American Bar Association and the Nixon Administration opposed it, despite the fact that it had originated with the Department of Justice. Assistant Attorney General William H. Rehnquist testified against key provisions in the bills at the Senate hearings. Both the Department of Justice and the federal judiciary were opposed to the "rigid and inflexible" time limits and to the sanction of dismissal of charges for noncompliance. This latter provision, Rehnquist argued, "is not only draconic, but quite one-sided. The result would be that a defendant held to answer for a serious crime would go scot-free—neither convicted nor acquitted."[8] As an alternative, the Department of Justice, with the concurrence of the Federal Judicial Conference, recommended that remedies for the problems of delay be handled by court rule, not legislation.

Opposition from the Department of Justice and the federal judiciary, coupled with the emergent Watergate scandal, put further consideration of speedy trial legislation on the congressional back burner. The judiciary used this time to mount its own alternative campaign for speedier criminal trials. In 1972, the Supreme Court, with the support and endorsement of the Judicial Conference, promulgated Rule 50(b) of the Federal Rules of Criminal Procedure. This rule provided in part that

> each district court shall conduct a continuing study of the administration of criminal justice [in its district] . . . and shall prepare a plan for the prompt disposition of criminal cases which shall include rules relating to time limits within which procedures prior to trial, the trial itself, and sentencing must take place, means of reporting the status of cases, and such other matters that are necessary or proper to minimize delay and facilitate the prompt disposition of such cases.[9]

The plan went on to authorize creation of a local planning and reviewing process of local district court judges who would develop and approve their own plan and then forward it to the national Judicial Conference for final approval.

To guide individual district courts, the Administrative Office of the United States Court prepared a model plan. It suggested that arraignment of those not released on bail be conducted within 20 days following indictment, and of those released, 30 days; that trial be commenced within 90 days after a plea of not guilty for those in custody, and within 180 days for those not in custody; and that a defendant be sentenced within 45 days of conviction.

The model plan identified a number of "excludable time" exceptions, including competency examinations, pretrial motions, interlocutory appeals, trials of other charges, motions under advisement, and continuances granted by the court. It also suggested a system of setting priorities in scheduling and outlined a number of procedural reforms. On the surface Rule 50(b) and the model plan are conscientious efforts to tackle a serious problem. The federal courts appear to have made a significant commitment to organize themselves more efficiently and to speed up their handling of criminal cases.

Although Rule 50(b) directed each district to draw up its own plans for the "prompt disposition of criminal cases," neither the rule nor the Administrative Office set any criteria by which delay could be gauged. Nor did they provide for any penalties in the event guidelines were not met. Most telling, a great many courts undertook no planning process and simply resubmitted the model plan as their own.

The nearly total ineffectiveness of Rule 50(b) was documented in a 1974 study by Andrew H. Cohn of the Yale Law School.[10] He found that plans prepared by ninety-two of the ninety-four district courts did little more than parrot the language of the model plan. They reported what their average case handling time had been in the past and made this their future target. Only one-third of the districts developed plans based on any significant independent assessment of their problems and needs. When they did depart from the language in the model plan, it was usually to substitute less restrictive language or to drop the plan's reference to the goal of "trial of all defendants within six months if practicable." And although Rule 50(b) and the plan explicitly sought to minimize the number of excludable time exceptions, the individual plans generally granted extensions of time limits. Some district courts provided that detained defendants were to be freed on their own recognizance, and others required that charges be dismissed if the court failed to meet its established time limits. But plans calling for strong sanctions for noncompliance usually allowed for generous time limits, and those plans imposing strict time limits generally had weak sanctions or numerous loopholes.

Nevertheless, officials of the federal judicial organizations staunchly maintained at congressional hearings that since Rule 50(b) had been adopted, the number of pending cases before the district courts had declined significantly and that cases were being disposed of more quickly.[11] However, a long parade of witnesses challenged these conclusions, and Professor Daniel Freed of the Yale Law School convincingly argued that the so-called targets in the district plans only served to reinforce old practices while pretending to bring about change.[12] Peter W. Rodino, Jr., chairman of the House Judiciary Committee, presented evidence showing the reduced backlog of pending criminal cases was due almost entirely to the decline in prosecutions under the Immigration and Selective Service Acts. Rodino summarized the sentiments of many witnesses when he criticized the federal judiciary for their failure to "put their own house in order." The problem, as he saw it, was that the courts' rule-making approach had no "spurs": it contained no sanctions to induce the courts to comply with established deadlines.[13]

District court planning efforts under Rule 50(b) were not uniformly discouraging. The judges in the Second Circuit, comprising New York, Connecticut, and Vermont, submitted detailed plans for reducing delay in their courts.[14] They were led by Chief Judge Irving F. Kaufman and Whitney North

Speedy Trial Rules and the Problems of Delay

Seymour, U.S. Attorney for the Southern District of New York.

But the problem of delay had been singled out by the judges in the circuit long before promulgation of the rule. Indeed, Rule 50(b) was similar to a local rule passed in the Southern District several years earlier, and the model plan was patterned after the District's own local plan.[15] If anything, the somewhat successful delay-reduction effort in the Second Circuit was the cause, not the effect, of Rule 50(b).[16] This exception suggests a disheartening lesson. Where the court rule approach was effective, a national effort was not needed, and where it was needed, it was not effective.

Keith Boyum, a political scientist, found that by themselves, new approaches (for example, master calendaring, individual calendaring, computerized case scheduling, rule adoption) had little effect in reducing delays. What did account for changes, he found, were the attitudes and *role-orientations* of the chief judges. Those who regarded delay reduction as an important objective and saw themselves as "administrators" achieved results whether or not they adopted new techniques or rules, while those who saw their position as largely honorific made few changes.[17] In the Second Circuit, long before promulgation of Rule 50(b), Judge Kaufman clearly had seen his role as an activist administrator and had targeted delay reduction as one of his objectives.

In 1973, Congress conducted extensive hearings on Rule 50(b) and concluded that the judges could not or would not act decisively on their own. In 1974, the Speedy Trial Act was passed over the objections of both the Department of Justice and the Federal Judicial Conference.

The Federal Speedy Trial Act of 1974

Passage of the Act. In 1973, Senator Sam Ervin and Congressman Peter W. Rodino, Jr., reconsidered the issue of court delay. Both were riding a wave of national popularity for their roles in the Watergate investigation. Ervin had conducted the televised hearings that had laid the groundwork for impeachment of President Nixon. Both men sought to capitalize on their temporary prominence by vigorously pressing their legislative proposal.

Like the 1971 proposals, the 1973 bill provided for strict time limits within which cases had to be moved from arrest to trial and provided for dismissal of charges in the event that these time limits were not met. Unlike earlier proposals, it provided for a five-year phase-in, during which period less restrictive time limits would apply and the sanction of dismissal would not be used. This provision stemmed from a recognition that the causes of delay were not fully understood, and that each district would experience unique problems implementing the act. Ervin's bill also included a research and planning mechanism. He saw the phase-in period as an opportunity for the judiciary, in collaboration with prosecutors, defense attorneys, court clerks, and others in each district to identify the specific problems of delay and to devise their own strategies to meet the requirements of the act.

However, this solicitude did little to appease the bill's opponents, who continued to argue that Rule 50(b) should be given more time to prove itself.[18] Judges complained that the bill's planning provisions encroached upon their management prerogatives,[19] and that requiring prosecutors, clerks, and other nonjudicial members of the criminal justice system to take part in planning violated separation of powers.[20] Even after passage of the act, judges continued to rail against it as unconstitutional.[21] Chief Justice Warren Burger called it "rigid" and went on to warn that it would "lead to unfortunate consequences."[22] A federal judge in Maryland held portions of the act to be unconstitutional infringements of the authority of the federal courts although the ruling was overturned.

The defense bar characterized it as a prosecution-oriented bill, one that would make it difficult for defense attorneys to handle their normal case loads. Defense attorneys complained that the bill's ten-day limit between indictment and arraignment was too short for negotiating a plea.[23] In Wisconsin, one prominent spokesman for the defense bar labeled it the "Speedy Conviction Act." This near unanimous opposition is understandable since, on the whole, delay benefits the defense. As a general rule, the defense bar will resist all efforts to expedite case handling since liberal continuance policies and lengthy delays allow them to manage their own time more effectively and increase their leverage in negotiations with prosecutors. By custom, courts have deferred to the interests of the defense bar, so it is hardly surprising to find such fierce resistance to policies that appear to upset such long-standing practices.

But neither were prosecutors enthusiastic. In 1971, the Department of Justice had adamantly opposed specific time limits and dismissal sanctions, and while this opposition softened after President Nixon's resignation, it did not end.

Thus, each major segment of the criminal justice community opposed the legislation. All were concerned that the time limits, the provision for dismissal of charges, and the requirement of a joint planning process would disrupt their time-honored ways of doing business. With but a few exceptions, they saw court rules as the appropriate way to solve problems of delay and favored Rule 50(b). But Ervin and Rodino pressed on, and Congress passed the Speedy Trial

Act of 1974 with its most controversial provisions intact.

Major Provisions of the Act. The Speedy Trial Act contains two sets of provisions, one dealing with time limits, the other prescribing the planning process. It established deadlines for the processing of criminal defendants in each stage from arrest to trial. Failure to meet its time limits results in dismissal of charges unless the factors causing delay come within its "exclusionary time" provisions. Congress also provided a four-year interim period, beginning July 1, 1975, for phasing in the speedy trial limits, with the law's strict time limits and dismissal sanctions taking full effect in July 1979. (The 1979 amendment to the act further delayed implementation until July 1981.) The phase-in period is closely tied to the act's requirement that planning groups be established in each district to assess local problems, identify additional needs, and develop plans for meeting the progressively stricter time limits during the interim stages of implementation. Congress sought to shape the contents of these plans by requiring that each plan include detailed descriptions of all new time-limit requirements, detailed plans for implementing these and related changes, and provisions for monitoring compliance by judges, prosecutors, and the defense bar.[24]

To many, the principal feature of the act was this planning provision. The act also assumed that the planning process was necessary to the implementation of strict time limits during the phase-in period. Nevertheless, extensive successful lobbying by the judiciary to extend the time limits and postpone implementation of the time-limit requirements casts doubts on the validity of much of the reporting data submitted by the courts. Congress found a pattern of court failure to comply with the strict reporting provisions of the act and feared that numerous dismissals might result if the July 1, 1979, date took effect.[25]

Speedy Trial Rules and the Problems of Delay

The act's time limits apply to the period between the arrest or service of summons or indictment and the commencement of trial and do not affect duration of trial, length of time to sentence, or the length of the appeal process. Time limits were established after a review of several national surveys and the testimony of a great many lawyers. The *maximum* limits are 30 days from arrest to indictment and 70 days from indictment to trial.[26] The interim time limits, which took effect on July 1, 1975, were considerably more generous (10 days between indictment and arraignment; 60 days between arrest and indictment; and 180 days from arraignment to trial), although each succeeding year they were decreased until the base time limits, given above, took full effect. Sanctions for noncompliance were to become operative one year later. Even with these compromises and delays, the results were considerably tougher than any previous state or federal speedy trial rule— judicial or legislative—hitherto adopted. They were, in fact, even more stringent than those proposed by the President's Crime Commission.

Though the act permitted dismissal either with or without the option of reinstating the charges, it anticipates that dismissal "with prejudice" (that is, with a presumption that charges cannot be reinstituted without good reason) should be the norm, and that dismissal "without prejudice" (that is, with the right to reinstate charges) would be the rare exception:

> In determining whether to dismiss the case with or without prejudice the court shall consider, among others, each of the following factors: the seriousness of the offense; the facts and circumstances of the case which led to the dismissal; and the impact of a reprosecution on the administration of this chapter and on the administration of justice.[27]

The wording of the statute is sufficiently ambiguous to allow judges who are unenthusiastic about the act to give greater weight to the former than to the latter considerations. And even before the final phase of the act took effect, there were indications that defense counsel could easily obtain continuances "in the interests of justice," and that dismissal "without prejudice" would be acceptable.[28]

While the act explicitly forbids exceptions to the time limits "because of general congestion of the court's calendar, or lack of diligent preparation, or failure to obtain available witnesses on the part of the attorney for the government,"[29] it goes on to catalogue a number of "excludable time" provisions, including the catchall exception: "if the *ends of justice* served by taking such action outweigh the best interest of the public and defendant in a speedy trial."[30] The act also permits exceptions in the event of a "judicial emergency." To qualify for this one-year extension, the district court must prove pressing court calendars and lack of resources.[31] Two additional exceptions to the "ends of justice" section are so broad and poorly tailored that they could permit placing the act's time limits in perpetual abeyance. They state:

> Whether the case is so unusual or so complex, due to the number of defendants, the nature of the prosecution, or the existence of novel questions of fact or law, that it is unreasonable to expect adequate preparation for pretrial proceedings or for the trial itself within the time limits established by this section;

> Whether the failure to grant such a continuance in a case which, taken as a whole, is not so unusual or so complex as to fall within clause (ii), would deny the defendant reasonable time to obtain counsel, would unreasonably deny the defendant or the Government continuity of counsel, or would deny counsel for the defendant or the attorney for the Government the reasonable time necessary for effective preparation, taking into account the exercise of due diligence.[32]

Speedy Trial Rules and the Problems of Delay

A possible exception not mentioned in the statute is the question of the defendant's right to waive the time-limit provisions of the act.* While the act explicitly recognizes waivers by the accused for other provisions, it remains silent on the right to waive the provisions entirely. At the time of this study, there have been no definitive rulings by appellate courts on this matter.[33] By way of comparison, state courts have almost uniformly been liberal in granting waivers, and the result has been, in effect, to nullify state time-limit requirements.

The act contains other ambiguities that might also be exploited. If a defendant is sick prior to trial, his attorney can claim the provision excluding "any period of delay resulting from the fact that the defendant is mentally incompetent or physically unable to stand trial"—because while the defendant is ill the attorney is unable to use the time to prepare for trial.[34] Another provision holding that delay "reasonably attributable to any period, not to exceed thirty days, during which any proceeding concerning the defendant is actually under advisement [is permitted]," might lead to routine thirty-day extensions.[35] Other specific exceptions also allow prosecutors to take more time for particularly pressing cases and to reindict defendants and start the clock all over again.[36] The act imposes restrictions and fines on attorneys who file frivolous motions or engage in delaying tactics, but they will probably not prevent lawyers from pursuing strategies they regard as beneficial.

The Initial Impact of the Act. Though the time limits and the sanctions are the most controversial sections of the Speedy Trial Act, its planning and research sections are its most important provisions. However, in the light of the additional two-year delay in implementation specified in the 1979 amend-

*The sole reference to waiver in the act appears in Section 3162(W)(2), which states: "Failure of the defendant to move for dismissal prior to trial or entry of a plea of guilty or nolo contendere shall constitute a waiver of the right to dismissal under this section."

ment and the casual attitudes toward the planning process, there is reason to be concerned.

Each district court was required to submit two plans to the Administrative Office of the United States Courts, one in June 1976 and one in June 1978. These plans were to contain extensive data on the district's interim efforts and a data-based list of needed resources.[37] They were to be formulated by a group of judges, court officials, representatives of the U.S. Attorney's office, members of the criminal defense bar, and an outside reporter. Three national judicial organizations shared responsibility for overseeing this planning process: the Administrative Office of the United States Courts, the Federal Judicial Conference, and the Judicial Conference Committee.

Unlike the explicit time-limits section of the act, the planning provision had to be left open-ended and vague.[38] This vagueness allowed each district to focus on its own distinctive problems, but it also allowed hostile judges the freedom to interpret their mandate narrowly and in effect to ignore the planning provisions.[39]

The extent and impact of this hostility cannot be underestimated. The federal judiciary had opposed the time limits, the planning provisions, the requirement to appoint a reporter, the requirement that the plans be submitted, and specification of the materials to be included in the plans. Once all these provisions were mandated, the federal judiciary continued to resist by minimizing their responsibilities and developing pro forma responses.[40] Even five years after passage, judicial hostility toward the act had not ceased. The Senate report on the 1979 amendment stated: "While the Administrative Office has demonstrated diligence and good faith in its efforts to guide the districts towards a reasonable application of the Act in practice, the Committee finds that, too often, the Administrative Office has erred on the side of caution."[41]

Speedy Trial Rules and the Problems of Delay

The Administrative Office of the Courts was required to develop a research guide for district court planning groups and to present a summary of all district court plans to Congress. The office developed a strict checklist type of outline and encouraged the districts to follow it in preparing their reports. It went on to suggest the use of standard language to identify the causes and consequences of delay and other problems. While this was convenient for the Administrative Office and the reporters, it discouraged extensive analysis by the planning groups.[42]

The office's reporting format also ignored a number of issues likely to be affected by the act: the effects of time limits on case preparation, scheduling difficulties confronting judges, administrative problems in the clerk's office, and, most important, the likely impact of the act on civil case loads.[43] An observer concluded that the net effect of these omissions was "to allow planning groups to comfortably skirt many of the issues which Congress had indicated that it wanted examined."[44]

Judicial resistance to the act has been a primary factor in delayed implementation. Planning groups have not collected adequate data on the impact of the act on civil case loads[45] or on the disposition of criminal cases.[46] Further, the Administrative Office has dragged its feet in providing information to the circuits about their responsibilities under the act.[47]

The judiciary had also opposed the statutory requirement that a reporter with administrative and research duties be appointed in each district, and once the requirement went into effect, it did little to facilitate the work of the reporters. The act required the reporters to probe deeply into the cloistered worlds of the federal judges, but more than one found it difficult to pursue his task.[48] The Administrative Office could have worked with the Federal Judicial Center on research and sent its experienced researchers out to the districts to help reporters.[49] Instead, it reflected the view of the chairman of the Judicial Conference Committee, Judge Alphonso Zirpoli, an outspoken critic of the act who earlier had observed it was "not . . . politic for either body (the Federal Judicial Center and the Administrative Office) to take an actively supportive stance toward planning."[50] This view was paralleled at the local level; in most districts, the "independent" reporters were little more than "assistants to judges charged with the responsibility of collecting data from the courts and the United States Attorney."[51] With but few exceptions, reporters did not undertake serious research, and planning groups rarely met—and even when they did, many members were absent.

It is not surprising that all this resulted in weak plans. The act's effects on different types of cases, the work pace of individual judges, the expected responses of defense attorneys and prosecutors, and other important factors were all but ignored.[52] Neither the first nor the second implementation report recommended changes in court administrative procedures.[53]

Of the many first-round plans reviewed in connection with this study, only the one prepared in 1976 by the Northern District of Illinois stands out as a detailed and thoughtful analysis. The report identifies many factors in delay: judges' health, their relative concern with substantive issues, efficiency of procedures, and so on,[54] but it offers no clear-cut proposals for improvement.[55] Despite its thoroughness, the plan for the Northern District was similar to the others in its conclusions: it called for repeal of the Speedy Trial Act.

With the exception of the Southern District of New York, the 1978 plans were also inadequate.[56] This is particularly surprising since the specter of sanctions was rapidly approaching.[57] A survey of the second set of plans submitted by the Ninth Circuit found that not one court anticipated any revisions.[58] Apparently, the plans were just another document the courts were required to file to an address in Washington. Congress found it necessary in 1979 to delay implementation of the time limits and to comment: "Too little has been learned about the potential impact of the Act's permanent provisions to predict with certainty that those time limits will be injurious to the Federal criminal justice system."[59]

Assessing the Speedy Trial Act. To date, neither the planning nor the time-limit provisions of the Speedy Trial Act appear to have been taken seriously by anyone but a handful of judges—who for the most part would have pursued these concerns in the absence of the legislation. Nor is there much basis for optimism; there are clear indications that the act's loopholes are being exploited to provide the appearance of compliance without any substantial changes. One federal judge who requested anonymity said, "We don't care if it's repealed or not; our court has figured out ways to deal with the act that don't cause us to change our practices at all. The act has caused us to be a little creative; that's all." These sentiments have been echoed by judges and prosecutors in other courts. "Excludable time" and "exceptions in the interests of justice" are terms being heard more frequently in the federal courts, and if this continues, the aims and purposes of the act will quickly be forgotten, with these formalisms all that remain.

Despite the hostility, foot-dragging, and the pro forma nature of most plans, the planning process has had important indirect effects. It has led some officials to reflect on ways to alter the administration of the federal courts and has led scholars to write about problems and offer solutions. It has also been an important impetus for modernization within the courts themselves; district court clerk's offices have been enlarged; administrative innovations have been adopted; scheduling practices have been tightened up; and grand jury sessions have been increased. New methods for tracking the flow of cases through the courts have been tried. Many courts now have a rudimentary "early warning" system to alert them of approaching time limits in individual cases.[60] The act has also accelerated long-standing plans to institute a computer-based management system in the federal courts. In order to reduce pressures upon judges, the functions and numbers of United States magistrates have been expanded since 1974. In the 1960s, a new profession, judicial administrator, emerged, and now a number of universities award advanced degrees in the field. The 1974 act has probably accelerated this trend.

Only a few of these innovations were required by the act, and most were planned long before its passage. Nevertheless, it is clear that the act has been a stimulus to their adoption and implementation. Since passage of the act, U.S. attorneys' offices have become more administration oriented as well. Prosecutors are now more conscious of the need to weigh the costs and benefits of prosecution and allocate their limited resources effectively. As a result of this selective process, criminal case filings in federal courts have declined markedly since passage of the act. But many of these benefits are illusory: they have been accomplished by transferring work to the already hard-pressed state courts.

The act has led to innovations for the defense as well. In a number of districts, prosecutors are now more accessible to defense counsel, more flexible in informal discovery, and more willing to engage in plea negotiations.* Many local court rules have been streamlined. Trial dates are now routinely set shortly after arraignment, and earlier pretrial conferences with magistrates determine whether a case will result in a trial or in some other disposition.

*This was the public position of the U.S. Attorney in the Western District of Wisconsin in 1980.

Speedy Trial Rules and the Problems of Delay

A study of the federal courts in New York City reported "significant complaints from defense counsel that judges were imposing deadlines that . . . caused grave and unnecessary problems for counsel and their clients."[61] Some allege that the act is forcing defense attorneys to decline representation of clients and to plead more of their clients guilty (although the increase in guilty pleas may be caused by the increased flexibility of prosecutors).[62] However, the act has shifted a balance of advantage to the prosecutor in one important way. Now prosecutors can rely on summonses to appear for an arraignment rather than seek arrest prior to indictment. This permits them to initiate criminal proceedings *after* they have prepared a case, thereby avoiding the thirty-day limit from arrest to indictment. To some, this practice contravenes the spirit of the act, which is to reduce all delay, not simply that which is included within its specified time frames. And the act can be criticized because it allows prosecutors to work on cases prior to arraignment, while it requires defense attorneys to prepare their cases within a shorter time.[63] More generally, this illustrates the ways officials can adapt to the new law without having to disrupt long-standing work habits.

Complying with the Time Limits. The success of the courts in meeting the time limits of the Speedy Trial Act is difficult to assess. Official reports show that in 1974 few districts met the limits, while by 1978 most did.[64] Other events have also occurred in the interim. The decline in disposition time parallels the decline in criminal case backlog. Between 1972 and 1981, the criminal case backlog in the federal courts was reduced by one-half. Some of this is due to new policies that shunt more cases to state courts, encourage earlier dispositions by offering more lenient pleas, grant magistrates greater authority to dispose of cases, and send some petty cases to pretrial diversion programs. Other important factors have been the end of the military draft and the selective service litigation it fostered and a revised policy on marijuana prosecutions. Also important is the changing age of the population: the "crime wave" of the 1960s and 1970s appears to have crested, at least for offenses that disproportionately involve young people. Still, if criminal case backlog has diminished, the civil case load is booming. The volume of civil cases pending trial reached a new high in 1978, and 10 percent of these cases were three or more years old.[65] The fear of an ever-rising civil volume as district courts scurried to satisfy the requirements of the Speedy Trial Act led many observers to be dubious of the ultimate value of this reform. In fact, one of the reasons given by Congress for prolonging implementation of the act was its unknown effects on the civil dockets.[66] Recent figures on federal court case loads suggest that this effect did not occur. Despite slight increases in criminal case filings in fiscal years 1980 and 1981 (after several years of declines), there was a decrease in the backlog of civil cases.[67]

Excludable time has become a strategic weapon used by the judges to fight the act. Early in the planning process one reporter noted: "The statistics in some districts indicate that the present level of compliance with the excludable time recording provisions of the Speedy Trial Act is so low that planning under the act is virtually impossible."[68] More recent observers echo this concern. Excludable time is compiled in an almost whimsical manner. For example, although virtually all criminal cases in the federal courts include at least one pretrial motion (an excludable time event), some districts never report this as excluded time while others always do.[69] But without such data, one reporter warned, "there can be no realistic attempt to achieve the planning functions of the act." This could "force the release of criminal defendants."[70]

Such problems may not materialize; the more likely response

to the threat of dismissals is adaptation. The planning groups were well aware of this, as is evidenced by the implicit suggestion in the planning report: *"greater familiarity with the act, particularly time considered excludable, will provide an even more favorable statistical performance* [in meeting the time limit requirements of the act]."[71] (Emphasis added.)

Formal compliance without actual change is widespread, and the vague "ends of justice" exclusion is among the means for achieving this sleight of hand. Of nine sample districts studied by the Department of Justice in 1979, one used this exclusion in 66 percent of all excludable incidents, while others never used it.[72] All, however, invoked one type of exception or another with some frequency. Thus, despite congressional efforts to force more rapid handling of cases, the act's "good cause" exceptions have quickly come to be loopholes, invoked at will.

Still, not all court officials oppose the act. I attended a day-long seminar that explained the intricacies of the act to defense attorneys in the Western District of Wisconsin. The judges, prosecutor, clerk, magistrate, reporter, and public defender were all supportive of the act. When I observed that this position appeared to be at odds with the views in other jurisdictions, the clerk acknowledged that the act would have virtually no effect in his district. Most criminal cases in the district, he reported, are relatively simple and straightforward and have always been disposed of well within time limits set by the act. At worst, he suggested, his court would have only two or three cases a year that might raise problems. In contrast, federal court in Chicago, where the judges are adamantly opposed to the act and have begun to routinely grant requests for exceptions, has hundreds of such cases. Thus, once again we find that the reform is embraced where it is not needed and resisted where it is.

State Efforts to Implement Speedy Trial Rules

The 1967 Supreme Court decision, *Klopfer* v. *North Carolina*, held that the Sixth Amendment's guarantee of a speedy trial applied to state as well as federal courts.[73] The Supreme Court's 1972 decision in *Barker* v. *Wingo* strongly suggested that speedy trial rules should emanate from legislatures rather than from the federal courts.[74] And since then a great many states have adopted new and seemingly stringent speedy trial rules. On the surface, it appears that there has been a revolution in state speedy trial policies. But these strongly worded statutes contain enormous loopholes. Either they provide for liberal exceptions, or, despite laudatory language, they contain lax time limits. Court decisions interpreting these rules have watered them down still further. Two recent surveys found absolutely *no correlation between the stringency of provisions in state speedy trial rules and actual disposition times.*[75] The state pattern is similar to that of the federal courts: expression of high-minded ideals coupled with symbolic responses.

Many states in search of speedy trial rules looked to the *Speedy Trial Standards* promulgated by the American Bar Association. Published in 1968, the *Standards* established time-limit guidelines for disposing of criminal cases, and it was the expressed hope of the ABA that they would serve as models for state courts and legislatures. To a large extent, these hopes have been realized; many states were quick to adopt the ABA *Standards* as their own.[76] Yet they themselves contain numerous loopholes, and state rules have often watered down their stringent features.

In most respects, the *Standards* are consistent with the Speedy Trial Act: they include overall time limits, require dismissal for noncompliance, grant priority to criminal over civil cases, and restrict excludable time. However, unlike the act, the *Standards* do not provide for mandatory time limits between indictment and commencement of trial, and most states have followed this less restrictive policy.

The *Standards,* state statutes, and court rules allow courts wide latitude in applying time limits. The *Standards* allow a continuance "only upon the showing of good cause and only for so long as is necessary" but do not define what constitutes "good cause."[77] The ABA defended this language on the grounds that it permitted "flexibility." But "flexibility" in speedy trial rules is what "all deliberate speed" in school desegregation was nearly thirty years ago—an excuse to do nothing.

A number of state speedy trial rules do not even approach the ABA's *Standards.* Some do not contain any specific time limits beyond requiring commencement of trial within a fixed number of "terms of court," a phrase that is meaningless since the advent of full-time courts.[78] But even the strictest speedy trial statutes reveal questionable commitment to reducing delay. Colorado's speedy trial statute is in "high conformance" with the ABA *Standards.* It mandates commencement of trial within 180 days of indictment but then provides for exceptions, one of which permits both the prosecutor and defense to delay a trial for an additional six months if time is needed to obtain more evidence or complete preparation.[79] Michigan has a 180-day time period for the commencement of trial as well as a dismissal sanction for noncompliance, but the appellate courts interpret noncompliance to mean that defendants must demonstrate that their cases have been prejudiced by the denial of a speedy trial.[80] In effect, the Michigan courts held that the state could deny a constitutional right if the possessor of that right could not show a specific harm in its denial.

Pennsylvania has had its speedy trial provisions eroded by state supreme court decisions, which is particularly surprising since that state's rule was promulgated by the high court itself.

Pennsylvania's Court Rule 1100 possessed numerous loopholes from the outset—prosecutors could petition for additional time upon a showing of "due diligence," and defendants could accede to continuances without giving reasons; but in *Commonwealth* v. *Mayfield,* the Pennsylvania Supreme Court went on to say that delay attributable to "congested dockets" was also a legitimate ground for granting extensions, a ruling that seriously erodes any incentive to overcome congested dockets.[81] In Montgomery County Common Pleas Court, prior to adoption of Rule 1100, extensions were routinely granted to the prosecution in up to 90 percent of all criminal cases. After adoption, extensions plummeted. In Philadelphia, a "fail-safe" system was instituted; all cases approaching the 180-day limit were automatically given preference.[82] While this had the effect of avoiding dismissals resulting from the time limit, it did nothing to reduce the overall time required to dispose of cases, and, in fact, it appears to have increased average disposition time. Without a definite commitment to reduce delay, time limitations may simply lead to a type of shell game in which the goal is to avoid dismissals. An article in the *Villanova Law Review* reported that in 1977 only ninety of the thousands of cases processed by the Philadelphia criminal courts were dismissed for violating the speedy trial rule.[83] "These statistics," the authors argued, demonstrate "that the efforts to obtain a speedy trial have been successful to an overwhelming degree."[84] These statistics can also suggest the opposite conclusion: the rule is having little impact on the courts, either because the court grants liberal exceptions to the time limits or because many cases are disposed of at the last minute. A more complete assessment would have to look beyond the dismissal rates and determine whether overall case processing time had been reduced without disturbing the requirements of due process, and whether the causes of delay had been diagnosed and treated.

A recent examination of the speedy trial provisions in Seattle showed that long delays in a handful of cases were eliminated only at the expense of an overall *increase* in time needed to process the great bulk of cases. The average disposition time of all cases was extended by one month.[85]

Additional evidence that speedy trial rules do not necessarily reduce overall time to disposition is provided in two studies of delay in state courts. The first, a study of ten states during the 1960s, revealed that there was no correlation between "strict" (that is, in conformance with the ABA *Standards*) speedy trial rules and swift disposal of cases.[86] Wisconsin, a "strict" rule state, disposed of no more cases as a percentage of new cases than Maryland or Connecticut, states of similar size that had only "vague" constitutional speedy trial provisions and no statutes. A study by Thomas Church found no correlation between speedy trial time limits and actual disposition times in specific cities.[87] Nor did it find any correlations between the strictness of speedy trial provisions, city size, case load, case complexity, and actual median processing time per case. In New Orleans, with a 730-day time-limit rule, the median number of days required for disposition was 67, while in Bronx County, New York, presumably governed by New York's "strict" 180-day rule, the median time was 343 days. Church concluded that differences were due to local attitudes and expectations—the local legal cultures shaped in large part by the defense bar.

In Pennsylvania, Church compared courts in Philadelphia and Pittsburgh, and in Texas he compared courts in Dallas and Houston. In Philadelphia, 28 percent of all cases exceeded rule-imposed time limits, while in Pittsburgh, only 9 percent did. In Dallas, 46 percent of all criminal cases exceeded the state's time limit, while in Houston the figure was 63 percent. Church found that in New Orleans, quick disposition of cases is valued by all elements of the bar and bench, while in the Bronx, attorneys "indicated real disbelief that a criminal case could or should move to disposition in less than a year (*six times* the average length of disposition in New Orleans)."[88]

Reflections on Speedy Trial Rules and the Problems of Delay

Delay: The Concept and the Problem. Delay is a blanket term covering a host of different problems caused by various factors, all requiring different responses. Delay is not one problem; it is a variety of problems.

How is the word used? At times delay refers to elapsed time between arrest and disposition (or some other stages in between). Delay at certain stages of the judicial process can mean numerous court continuances. A large backlog holds up disposition of cases, as does courtroom congestion and confusion. "Dead time," when judges, jurors, or attorneys are present in court and waiting for a trial to start, is one type of delay. Poor notification of witnesses, resulting in needless court appearances or missed appearances, causes yet another kind of delay. In short, the word *delay* signals frustration over a variety of apparent inefficiencies in the court. But because it means so much, it is in danger of obscuring rather than clarifying issues. As the authors of one very detailed study by a group from the American Judicature Society observed, "Two years of research has convinced us that simply viewing delay as one problem is not very helpful. . . . We prefer to view delay as a symptom of other problems that exist within a court system." In locating the sources of these various problems, the study further concludes, "Our research indicates that *all* actors may be responsible for unnecessary case processing time."[89]

Some have suggested that heavy case loads and delay cause judicial lenience—that they force prosecutors to plea bargain more casually and judges to offer heavily discounted sentences. Yet evidence suggests that there is little if any correlation between length of time to disposition, or courtroom "congestion," and severity of sentence. Holding constant seriousness of charge and nature of incident, cases that move to disposition quickly and those that take longer are equally likely to lead to the same types of sentences.[90] Studies by the National Center for State Courts and the American Bar Foundation have found no clear relationship between size of docket and length of time to disposition.[91] Nor have these and other studies found any relationship between the judges' work loads and length of time to disposition.[92] Disturbances in jails have little to do with length of pretrial delay.[93] Finally, increased courtroom personnel and resources—seemingly the obvious remedy—do not reduce delays. Indeed, a marked increase in resources can lead to a slower pace of doing business by fostering an increase in trials.[94]

Multiple Causes for Various Types of Delay. Despite their shortcomings, speedy trial rules represent an important step forward and sometimes show a sophisticated understanding of the problems of delay. They do not fully appreciate the complexity of the factors giving rise to delay. Setting time limits for the various stages of the criminal process establishes hurdles through which attorneys must jump but does nothing to alter their incentives to delay. And threatening dismissal for noncompliance may even enhance the use of delay as a strategy since as a group it is defense attorneys who have the most to gain through prolonging their cases.[95]

The pace and manner of handling cases are part of the fragile balance of the courts. To alter them will set up a chain reaction throughout the entire system and precipitate new problems. Or it will lead to a hydrauliclike adjustment, which accommodates to the change by adaptive behavior. These are reasons to proceed cautiously and incrementally. But this is just what most speedy trial reforms—and especially those at the state level—have not done. A problem-specific focus, I suspect, is likely to lead to a more complete understanding of the problems and a more meaningful, but less publicly appealing, set of remedies.

Martin Levin has found that lengthening time to disposition often saves time in courts and in preparation, more important "savings" than earlier disposition.[96] Levin reports that this type of delay is also functional for defendants, who need time to reconcile themselves to their not very pleasant futures. It is often viewed as a necessity by private attorneys who are reluctant to dispose of a case without first having been paid for their services.

Speedy trial rules do not necessarily provide incentives to reduce the number of continuances. One judge observed that "any defense attorney worth his salt will use whatever tactics within his means to further the interests of his client, and if this means numerous continuances, he'll waive the speedy trial provisions." An examination of practices in "strict rule" states found that this judge is correct: few defense attorneys attempt to work within the time limits.[97]

A Chicago study found that 70 percent of the unexplained continuances requested by private attorneys appear to be related to efforts to boost their fees.[98] Attorneys often use continuances and the filing of perfunctory motions to give the appearance of action. They hope that such delays will wear down the prosecution witnesses. As Macklin Fleming has observed, a continuing cause of delay is overcommitment by attorneys who have a tendency never to say no to a lucrative case.[99] Speedy trial rules do little to affect any of these incentives.

Finally, we must consider the congestion and confusion in the courthouse, something readily observable to anyone who has set foot in a court. A judge in one courtroom may be swamped with a dozen arraignments, while down the corridor another sits in his chambers reading because his scheduled trial folded at the last minute. In one room, jurors may wait for a wrangle before the bench to conclude, while in another, every third or fourth case may be continued or dismissed because a key witness is not present.

Speedy trial rules focus on only one type of delay, the reduction of elapsed time. But, as has been suggested, even if they are effective, there is no assurance that the other problems of delay would be ameliorated, and in fact they might even be increased as a consequence. Imposing time limits does little to tackle the various underlying causes of delay of all sorts, nor does it do anything substantial to affect the incentives of court officials, a step that appears to be indispensable to devising effective management improvements.[100]

Speedy trial rules have generated more controversy than action. Most are amorphous enough to facilitate adaptation instead of real change. The response by the federal courts most clearly demonstrates this. Resisting the rules from the outset, the federal judiciary finally adopted a weak and ineffective rule to ward off congressional action, and when this failed, the judges continued to resist by perfunctory participation in the mandated planning process and by liberal interpretation of the time limits.

At worst, speedy trial rules are little more than symbolic responses. At best, they suffer from trying to do too much—offering a single, simple solution for what is in fact an extremely complex set of problems. In either case, the promulgation of rules and time limits has as of yet contributed very little toward reducing delay.

Strengths of the Speedy Trial Movement. The primary effects of the speedy trial movement have been indirect. But this in itself is significant. The Speedy Trial Act represents a more sophisticated and far-reaching approach to delay than has been typical of other organized efforts to cope with case load. By providing for a lengthy planning and implementation process, Congress in effect recognized that the problem of delay was complicated. As Professor Daniel J. Freed, one of the principal architects of the act, has said, it is primarily a "planning and resource" provision. To the extent that the federal system serves as a model for the states, this perspective will gain wider adherence and foster still more sophisticated thinking about court reform. Thus, here, as with the other reforms, the general momentum of concern may be more important than any specific provisions.

Conclusions

What are the lessons learned from this examination of efforts to formulate and implement speedy trial rules? First, delay is not the well-defined problem so many think it is. The various types of delay serve many different functions for the courts and should be considered in their various component parts. Trying to reduce overall time to disposition may be too broad a target. Focusing on specific problems rather than on delay in general may be appropriate. For instance, the causes and effects of delay in cases where defendants are unable to post bond are quite different from those in cases where defendants are out on bond, perhaps for long periods before being convicted and sentenced. But in both cases, specific moves are needed to deal with targeted populations. The former requires increased pre-trial release and due care by defense attorneys, while the latter may require special prosecution units focusing on major offenses. Similarly, cases involving charges of violence or career criminals may reasonably be regarded with greater urgency than those involving less serious issues.

Above all, it must be recognized that delay will always be used strategically by attorneys, and that any new rules or provisions will also be put to strategic use. This is the nature of the adversary process. Finally, we must simply accept the fact that courts will always be "inefficient." Cases will always fold at the last minute, jurors will always have to sit idle and be on call, and attorneys and witnesses will always have to wait.

However, incrementalism and problem-specific solutions are likely to be fruitful, even though they will not be seen as "bold" or "new." The planning provisions in the Speedy Trial Act may give rise to some limited but effective reforms. A number of modest efforts targeting specific types of delay have been adopted in recent years. Operating in the shadows of the more visible and dramatic speedy trial rules, they are in fact much more likely to be effective in solving some of the specific problems of delay. There has been a slow but marked evolution in how judges perceive their roles. Traditionally, judges have regarded themselves as passive referees, but increasingly they are adopting an active stance, scrutinizing reasons for continuances, limiting issues, controlling their calendars, and questioning jurors. All this is reinforced by a growing concern with judicial administration, leading to the appointment of full-time administrators, the use of computerized information systems, and the adoption of other management devices.

Concerns with efficiency have given rise to other specific

problem-focused innovations. In New York City, the Vera Institute has pioneered in projects that might contribute significantly to overcoming some of the most frustrating features of delay and confusion in the court, in particular, non-appearances by witnesses. It developed a victim-witness notification program to remind witnesses of scheduled court appearances. This program, now operating in a number of cities, also identifies complainants who will not show up in court and do not wish to continue the process. Early notification of the court of their intentions saves preparation time. The Vera Institute has also helped prosecutors in the Bronx and Brooklyn develop major offense bureaus, which handle cases involving especially serious charges or known prior offenders. Similarly, across the country programs have been developed to better utilize jurors, handle cases of those unable to post bond, and expedite motions practices. While none of these reforms tackles the problem of delay directly, all address significant problems underlying the general concern with delay and focus on specific problems so often glossed over in the general concern with delay.

Still, to the extent that the right to a speedy trial is in fact defendant-based, allegations of its denial should properly be pursued by adversarial proceeding as well as legislative and administrative means. To date, appellate courts have been derelict in their responsibilities to protect this constitutional right. But with this marked exception, incremental and problem-specific approaches have accelerated in recent years and, while unheralded, are likely to accomplish more than across-the-board speedy trial rules can accomplish.

Part 3

Assessment

Chapter 6

IMPEDIMENTS TO CHANGE

Impediments to Thinking about Change

When I asked a defense attorney in New Haven about his reluctance to litigate obviously unconstitutional practices of that city's courts, he stated: "My effectiveness in individual cases depends on how well I can take advantage of the chaos in the system, and, above all, I must keep the interests of my client in mind." In short, a limited perspective is functional, and broader considerations are not. Chaos benefits someone—perhaps everyone—in the court system.

The attorney's view supports Raymond T. Nimmer's observations about the desultory history of court reform. Nimmer writes:

> In most instances, not only is there no general desire to change, but there is a systemic tendency to retain the *status quo*. . . . Prior judicial practice is not arbitrary but reflects an accommodation of the interests of participants. In the abstract, this accommodation may not be ideal, but in context, it is at least acceptable to participants. . . . While a reform may affect one or several factors, other factors will often be unaffected. These will tend to perpetuate prior practice, thus enhancing the overall systemic resistance to change.[1]

Nimmer could have gone further. The central obstacle to change in the courts is not the resistance to reform, but is, more fundamentally, the lack of interest in even thinking about change.

This is not to suggest that there are no efforts at *planned* change—indeed, this book has explored quite a few—only that there is little incentive for those engaged in day-to-day administration of the criminal courts to think about systemwide changes or, when they do, to pursue them vigorously. But when change comes, as we have seen, it is often initiated by "dramatic events" and offered as a "bold solution" that is promoted as a panacea. Such conditions do not give rise to serious thinking or realistic expectations. The several components of this dilemma are spelled out below.

Problems of Crisis Thinking. Planned change in complicated and fragmented organizations emerges from slow, partial efforts. Yet because of the salience of the issues, in the criminal courts bold crusades are undertaken against little-understood enemies, often fanned by an atmosphere of crisis. It is tempting for reformers to cut through complexities, point out enemies, and offer bold strategies. In order to mobilize public support, reformers must often offer dramatic plans that are both vague and simple. But these very strategies that facilitate innovation undercut implementation.

Lack of Historical Perspective. Crisis thinking lacks historical perspective. It also drastically underestimates the tensions and interdependence in the criminal justice system. A common problem, we saw, was that proponents of change tend to view the problems they tackle as unprecedented and plan their enterprises as if others in the court were equally enthusiastic about them. A historical perspective shows that many problems have long histories and stem from deep-seated and insolu-

192

ble tensions, and that the typical stance of others is to resist and adapt, not to embrace reforms.[2]

The Inevitability of Crisis Thinking. Reformers are not all shortsighted or careless. They must achieve change in the face of official resistance and public indifference, and often they see a crusade as their only means. But while a crusade may be successful in mobilizing support, it must of necessity overstate the problem and oversimplify the remedy. This, reformers acknowledge, is the price one pays for pursuing a difficult goal. Unfortunately, it often is too high; the crusade becomes an end in itself, or the virtue of simplicity is purchased at the price of understanding.

Politicians are aware that public support and career advancement are more likely to be gained by pointing to an enemy, preferably a nonreactive one, such as case load or crowding, and initiating a bold new campaign against it than by working quietly to implement incremental changes in existing institutions. Thus, it is not altogether unanticipated or irrational that dramatic and expensive new programs fail; indeed, to reformers, many crusades are not failures. Symbols can succeed while actual policies fail.

Unfortunately, the Law Enforcement Assistance Administration (LEAA) exacerbated this problem: Created by Congress in 1968, LEAA was charged with fostering comprehensive planning in what was acknowledged to be a highly fragmented criminal justice system. Congress also made available through LEAA millions of dollars for innovative programs and created a national research institute with specific evaluation duties. LEAA could have quietly experimented with various approaches to pressing problems, but the temper of the times prohibited this. At the national level, LEAA was used to promote the Nixon administration's image, and at the state level its programs were dominated by the very groups it was seeking to change. The idea of developing and testing new ideas through research, experimentation, and evaluation gave way to the hoopla of dramatic displays, the development of elaborate crime-fighting equipment, rhetorical excess, a continual search for purpose, and business as usual.[3]

Perhaps most disappointing was LEAA's inability or unwillingness to foster research and development and careful evaluation, especially in the area of pretrial diversion. Once it had tagged diversion as a "good idea," LEAA was able to induce literally dozens of cities to fall into line, although it was fully ten years before LEAA paused long enough to support the first careful evaluation of the idea.

The Fallacy of Formalism. Reliance on formal description of the criminal justice process as a basis for diagnosing problems and constructing remedies can be termed the fallacy of formalism. The problem, of course, is that formal descriptions do not adequately represent actual practices.[4] At best, they oversimplify; often, they are wrong. And however appealing in the abstract, principles look quite different in practice.

Yet reformers find it easier to marshal support behind appealing principles than to explain the murky details of actual administration. Inevitably, public communication and the limits of legal language require simplification, but it can easily degenerate into a symbolic politics that divorces discussion of policies from actual practices.[5] Proponents of determinate sentences appeal to the equity of the idea but neglect to consider that offense labels can mask significant variations in actual seriousness, or that structured sentences can enhance prosecutors' incentives to plea bargain. We saw signs of this in the examination of various sentence reforms. Advocates of pretrial diversion argued that diversion would provide a quick and informal alternative to time-consuming adjudication—as if cases in lower courts really do go to trial in large numbers. Indeed, we saw that one reason why so many diversion programs did not attract clients was that the supposedly formal and harsh court in fact offered a speedier and more lenient alternative than diversion programs. Supporters of delay reduction often act as if court delay is a discrete problem to be overcome by expanding court resources, without firmly attacking the incentives of defense attorneys or realizing that such expansion invites still more cases from a nearly inexhaustible supply of controversies. In all these cases, retreat to a formal characterization of the criminal process seriously misrepresented reality, led to faulty diagnoses, and contributed to policy failure.

The use of obfuscating language by reform activists themselves is a variation of this fallacy of formalism. They meet with one another to offer "technical assistance," attend "training sessions," and make "site visits." While such terms must be used as verbal shorthand from time to time, their frequent use fosters a banality of discourse that provides a false sense of precision and diverts attention away from concrete problems.

The pragmatist movement in the early part of this century taught us to distrust formalism and reification, to insist that principles be examined only in relation to concrete settings, and to base reform proposals on the logic and incentives of those whose practices they seek to change. This mood leads, in the words of Edmund Cahn, to a consumer perspective on reform, one that focuses on social problems from the point of view of those who actually experience them.[6] This perspective is more concerned with understanding the sense of injustice that arises in concrete settings and how to reduce it than with fashioning a comprehensive theory of the just policy. It fosters an experimental approach to change, tentatively moving away from what is inadequate in the hope of making incremental improvement.

Outsiders and Change

Given the lack of incentives for systemwide changes within the courts, it is not surprising that innovation should often come from outsiders. Thus, another dilemma: those who are in the best position to assess the needs of the courts have the least incentive to innovate, while those who have the incentive do not have the detailed knowledge.

If change is initiated from within one part of the court, it is likely to affect the internal operations of that agency and only indirectly the whole system. A district attorney may shift some of his assistants from trial work to screening of cases, or from appellate to trial work, or he may convert to a computerized record-keeping system. Such changes are likely to have only marginal effect on other court operations. But if a single agency unilaterally implements a new policy that has systemwide impact, then it is likely to be greeted with resistance and adaptation.

In the long run, two factors reinforce each other and contribute to the lack of innovation in the criminal courts. First is the need for the various officials—even though nominal adversaries—to cooperate so that courtroom behavior is predictable. Second—because the courts possess hydraulic qualities in which each component can effectively thwart changes—is the lack of incentive to try to change.

If criminal justice officials have few incentives to initiate systemwide innovations, then who does? People and agencies outside the criminal courts: legislatures, appellate courts, entrepreneurs in nonprofit service delivery organizations, special commissions, and, more recently, LEAA. Although nothing has prohibited local judges and prosecutors from developing their own guidelines for sentencing, for the most part, the major efforts at sentencing reform have been pursued by legislators.

Impediments to Change

Even the appellate courts are not particularly close to the problems they so often must involve themselves with. Appellate court judges, as Donald Horowitz has pointed out, must rely on artificial and frozen records for their information, possess few information-gathering resources of their own, and usually are not able to consider effectively the impact of policy making on matters of broad social concern.[7] Perhaps sensing their limitations and the growing public criticism of activist judges, the federal courts have become increasingly reluctant to serve as sources of change. Indeed, despite specific constitutional provisions, the United States Supreme Court has ducked the issues of bail and speedy trial.

While outsiders may be able to transcend the limited perspectives and incentives of those who work daily in the criminal courts, their remoteness from the courts prevents them from understanding the byzantine realities of the criminal justice process, and, as a result, their efforts are often misdirected. Furthermore, they rarely have a continuing concern with the problems of the criminal courts. They show concern only when they are forced to deal with a problem, with the electorate at large, or with some other constituency. Indeed, success to many outsiders means adoption, like passing a new law or announcing a new ruling. Continuing interest and the authority to deal with the many factors that can subvert new policies are needed.[8]

Problems of Implementation

After changes are initiated, they must be implemented. Promises must be translated into actions, and different people—usually with their own agendas—enter the picture. The complexity of joint actions and multiple perspectives makes implementation of even the most simple public effort incredibly difficult. Two prominent students of public policy, Jeffrey Pressman and Aaron Wildavsky, suggest that we turn our standard concern on its head and ask not why do programs fail, but why it is that a few of them are able to succeed.[9]

Several features of the courts exacerbate the normal tendency to failure. First, the fragmentation of the criminal justice system facilitates judgments of success even as reforms fail. Second, many reforms have sought to circumvent the sluggish institutions by creating new programs, but these quickly became part of the problem. Third, success of programs has often been declared prematurely.

Fragmentation. What one sees and how one assesses it depends upon where one sits, and in the criminal justice system there are many seats. Thus, for years the claims of success by pretrial release agencies and pretrial diversion programs went unchallenged by prosecutors and defense attorneys, either because they seemed plausible or because no one cared to scrutinize them carefully.

Even when reforms are mandatory, they may not have their intended impact. The adaptive and perfunctory responses to compulsory change were seen in the judges' responses to Michigan's firearms law and by the federal judges' response to the rule requiring speedy trial plans to be filed. Similarly, despite the aim of the new California sentencing law to reduce sentence disparity, there is mounting evidence to suggest that the problem remains intact. A more general problem is value conflict. A great many judges quite simply do not regard liberalized pretrial release or mandatory sentences as desirable and thwart their implementation. New policies emphasizing one set of values do not neutralize long-held views to the contrary. Yet, to be successful, reformers must ultimately alter the incentives

Impediments to Change

of those whose behavior they seek to change. At a minimum this requires that they first understand contrary points of view rather than follow the natural impulse to ignore or dismiss those who differ with them. It also implies that they attend to the mundane details of implementation at the lowest levels of organization when their propensity is to rest content with vague pronouncements by leaders.

Newness. Each of the reforms we have examined in this book—bail, diversion, sentencing, and speedy trial—was premised on the belief that a new institution or set of rules was required. Special bail and diversion agencies were created because existing institutions were presumably unwilling or unable to meet their obligations. New rules seemed necessary to cope with long-standing problems of sentencing and delay. The attraction of *new* institutions is undeniable, but new institutions lead precarious lives. They may not be able to deal effectively with entrenched powers. They may not possess political patrons to run interference for them. They may be rooted in false premises about problems. They often have little power to alter incentives of entrenched officials who support old practices. And while a new, highly charged nonprofit corporation providing services on a contract basis may not have to put up with the headaches of seniority, civil service, and high salaries, neither does it benefit from broad-based political support and secure personnel. Most debilitating is the fact that new organizations age rapidly; what is new one day is old and established the next. In a system marked by fragmentation and hypertrophy, the strategy of creating new institutions rather than invigorating old ones may be counterproductive. The widespread desire of reformers to circumvent rather than resuscitate sagging probation agencies is, I think, a sad case in point.

Premature Judgment. A new idea might work well at first.

But initial success does not assure lasting success, and old patterns may reemerge. Industrial psychologists have a name for this phenomenon: the Hawthorne effect. The term refers to a 1930s productivity study at the Western Electric Company's Hawthorne, Massachusetts, factory. In that experiment, researchers found that after the introduction of innovations on the assembly line, employee productivity increased dramatically. Later research suggested that the improvements were due more to the increased attention paid employees during the study than to any actual changes in work arrangements.

Nevertheless, it is usually only "experiments" that are evaluated, written about, and publicized. Years later, bail reformers still return to the articles alleging the dramatic success of the Manhattan Bail Project to justify their own release programs. Similarly, proponents of pretrial diversion return to its early laudatory studies. The findings of this book, however, suggest that the Hawthorne effect is at least as applicable to the criminal courts as it is to industry. The result is premature judgment about impact and inadequate attention to problems of institutionalization. Ultimately, widespread implementation requires that experimental efforts promoted by dynamic entrepreneurs be administered by ordinary people. Innovators and evaluators typically fail to address these issues.

Problems in Routinizing Reforms

It is rare to find an innovation that is carefully initiated and even rarer to see one successfully implemented. But it is rarer still to find a workable new idea well institutionalized.

Usually initiated by outsiders—private foundations or LEAA—diversion and bail agencies were initially quite successful, but once they were institutionalized they faced one crisis after another. Hundreds of such programs across the country have been started with federal funds, only to wither for lack of support once those funds were withdrawn. Even those that have survived often remain in financial difficulty or stay alive by taking on different responsibilities. Programs in New York, New Haven, and California had histories of continuing financial crises brought on, in part, by initially unrealistic funding.

While innovations may be adhered to at the outset, once financial reality has set in and the glare of publicity has declined, there is great incentive to revert to old practices. In Michigan, this adjustment was almost instantaneous and total. In New York, the adjustment was slower but even more complete. In Boston, the courts were able to absorb an increase in work because of the general decline in the overall case load and the relatively few cases involved. At the time of this writing, the California prisons are beginning to reach the bursting point, and a variety of adaptations are beginning to surface. The Federal Speedy Trial Act has been watered down with amendments even before it goes into full effect.

Successful innovators are rarely successful administrators. New programs experience a rapid loss of moral fervor: charismatic spokespeople are replaced by bureaucrats; prestigious sponsors move on to other things; young and enthusiastic staffs age and become more security conscious; co-optation and adaptation become necessary for survival. Concern for original goals gives way to concern for organizational maintenance and the program objectives of the new generation of administrators.

Nowhere were these problems more vividly demonstrated than in the pretrial release efforts in New York City. Once Vera's ROR project had demonstrated its worth, it was transferred to the Office of Probation. But the transfer led to a shift in goals and a dramatic decline in effectiveness.

In San Jose, the diversion program significantly altered its goals when LEAA funds were exhausted. In New York, once Rockefeller was out of office, his drug law was repealed. In Michigan, the mandatory firearms law and no-plea-bargaining policy of the district attorney were never taken seriously by Detroit judges. The long-term effects of the California and Massachusetts laws remain to be seen.

Problems of Evaluation

If the planned changes examined in this book were as ineffective as I have argued, why were they supported for so long? Why, during a period in which the watchword in federal projects was *evaluation,* did so many problems go unnoticed?

The answers again are lack of incentive, fragmentation, and multiple perspectives. Proponents of reform have little incentive to evaluate; they *know* their ideas are good. For many, success is defined by the ability to adopt, not implement, a new idea. To lawyers and legislators, success is often "winning the case," passing the bill, or handing down the proper decision.[10] To foundations, it is often securing a good press or involving prominent people. Thus we need not lapse into cynicism when we find that for many officials the problems of delay were meaningfully addressed by Rule 50(b), or that the problems of sentencing disparity were overcome with the adoption of determinate sentencing.

To evaluate is to clarify, to unmask, to set standards. Administrators fear evaluation, a process that, if pursued honestly, must either hold programs to their promises or reveal unpleasant realities.

Even when there is a desire for evaluation, fragmentation

and the agency perspective are likely to limit its scope. LEAA-mandated state planning agencies, legislatures, local coordinating councils, and private agencies all profess interest in systemwide change; yet they tend to ignore the systemwide impact of change as well as evaluation. Here, too, more public support and prestige will attach to announcements of bold initiatives, exciting successes, and the promises of leaders than to the upgrading of routine practices and reports of mixed results on the activities of low-level subordinates and discussions of the mundane details that affect actual administration on a daily basis.

This preference for the bold and easily understandable is not served by many "scientific" evaluations. The better evaluations are, the less likely they are to be clear-cut, newsworthy, or even helpful. Furthermore, the causes of observed changes are complex. The evaluations of the impact of the Massachusetts gun law in Boston found that there had been some changes after passage of the law, but they could not (and did not) easily attribute all these changes to the law. Similarly, the emerging assessments of the new California law are tentative and inconclusive.[11]

The more rigorous an evaluation is, the more likely it is to sound inconclusive. It is by now a cliché that most evaluations end with a plea for more research. But what is sound practice for the researcher is seen as obfuscation by the policy maker, who wants simple yes or no answers.

Furthermore, good evaluations are expensive and time-consuming. They often are not complete or timely enough to be of use in the formulation of policies. Legislative and budgetary agendas do not wait for the results of evaluations; one of the most controversial provisions in the Rockefeller drug law was repealed before the results of the evaluation were formally reported. Similarly, the decision to restructure the New York Court Employment Program was made before the final results of the elaborate evaluation were known.

New programs are subject to unanticipated obstacles that can retard or derail them. Both programs and evaluations must be flexible; but this flexibility in turn facilitates manipulation and distortion. Recall the early favorable evaluations of pretrial release and diversion programs and the study on the effectiveness of Rule 50(b) by the Administrative Office of the Courts. The shortcomings of these studies were not obvious to non-researchers.

Evaluators are often co-opted by those they study. They are reluctant to bite the hand that feeds them. But even well-meaning and conscientious evaluators can come to share the same enthusiasms and moral commitments as project staff and, as a result, lose objectivity. Or they may get bogged down in concern for the administration of the project and come to define success in terms of smooth operations rather than ultimate impact.[12] Thus, it is wise to separate project evaluation from administration, a suggestion that is reinforced by the fact that most of the best evaluations reviewed in this book were undertaken by investigators with some distance.

Still, distance is purchased at a price, as was seen in the early efforts by the Labor Department to evaluate the pretrial diversion programs it funded. Although these programs initially agreed to cooperate with evaluators, once they had received their funds, they had little incentive to do so.

Finally, standard evaluation designs can be a hindrance to meaningful program assessment if they emphasize easy-to-quantify factors at the expense of other qualitative factors. Many evaluations fail to place projects in perspective and recognize the importance of personality. For instance, the failure of the diversion programs officials to cooperate with the evaluators selected by the Department of Labor should itself have become an important focus of the evaluators. During the early stages of project initiation and implementation, evaluations that focus on process may be more valuable than those focusing on impact and outcome.

Conclusions

Scholars are finding that many innovative programs fail in their implementation.[13] This book suggests that the picture is bleaker: the causes of failure are found at every stage of planned change. Often, failure is rooted in conception, in a fundamental misunderstanding of the nature of the problem, the dynamics of the system, the nature of the change process, and attention to detail at the service delivery level.

The central and continuing obstacles to change in the criminal justice system are fragmentation and adaptation, and there are two approaches to coping with them. We can seek increased coordination, or we can devise a strategy that takes these conditions into account as immovable objects around which reforms must be built. The dominant approach taken in reforms examined in this book has been the former, to seek improvements through greater coordination and better management. This approach can be called *administrative*. Administrative changes try to impose a bureaucratic form on an inherently antagonistic adversarial system. The proponents of administrative change neither adequately consider the actual operations of the courts nor appreciate the theory of the adversary system.

Perhaps the greatest danger in the administrative strategy is that it will work, that it will transform a contentious and embattled group of professionals into cooperative bureaucrats.

Pretrial release agencies are increasingly turning their attention to the goal of predicting "dangerousness" and, in doing so, legitimizing the concept of preventive detention. There is some justifiable concern that the dramatic increase in the use of court administrators will emphasize judicial productivity at the expense of traditional adversarial concerns.

In the short run, administrative reforms may appear successful, but once institutionalized, they can easily become part of the fragmentation that is the source of so many problems. The administrative strategy is questionable for still another reason: it blunts a "rights strategy" for change. This latter strategy is more consistent with the adversary system because it seeks to translate interests into legally guaranteed rights, and because it accepts and builds on the fragmentation and combativeness inherent in the adversary system. It provides functions compatible with form. Perhaps the single greatest change in the operations of the criminal courts in the past half century has been the expansion of the right to counsel. Not only has it done the obvious—provided protection for the accused—it has led to improvements in the quality of the work of police, prosecutors, and judges. But at the same time, it may have contributed to the appearance of a decline. Expanding rights in the criminal courts raises expectations, exposes practices to closer scrutiny, and makes decisions more contentious and cumbersome. Thus, an irony: as things get better they appear to get worse.

While rights-based reforms have contributed to significant improvements in many areas, the administrative strategy has blunted efforts in others. Professor Foote's hoped-for constitutional revolution in bail failed to materialize for a great many reasons, including the rise of "scientific" administration of pretrial release. This approach, in turn, transformed what might have been the development of a set of rules and rights into a set of administrative axioms—"scientifically" derived predictors of appearance and nonappearance—and in the process translated interests into privileges to be dispensed at the discretionary judgment of "experts." If the rise of bail agencies, diversion programs, and speedy trial provisions—all offered in the name of pursuing the interests of the accused—has failed to make significant strides in achieving their goals, there is little reason to believe that other administrative reforms will be any more successful.

A rights strategy is not an end in itself; its primary virtue is that it *can* focus on specific problems and is compatible with an adversarial process. Thus, it stands in contrast to some of the more sweeping policies examined in this book. The broader strategy is one that is pragmatic and problem oriented. This, as we will see in the next chapter, can take a variety of forms.

Chapter 7

TOWARD A STRATEGY

FOR CHANGE

Benefits of Failed Efforts

We have found that important changes often occur unheralded as by-products of seemingly unsuccessful efforts or as the consequence of general social change.

Consider, for example, the problem of pretrial release. Over the past two decades, practices have changed dramatically, even though the changes are due more to increased general sensitivity than to the development of any specific new programs. The result? Pretrial release agencies that have not worked, but changes in pretrial release policies that have. Diversion, delay reduction, and sentencing efforts have experienced similar patterns. Many of these new efforts have come to serve other related functions. Release agencies track down nonappearants and notify witnesses. Determinate sentencing reform in California has not only increased interest in harsher sentences but has also stimulated concern with sentencing disparities and discrimination in a great many other states. The existence of diversion programs calls into question the wisdom of incarcerating nonviolent petty offenders and perhaps has increased the legitimacy of alternatives to conviction and incarceration. Thus, although new institutions have not yielded their expected benefits, they can serve to keep serious problems on a fickle national agenda. They stand as symbols of a public commitment to change. This helps explain why so many programs can be judged unsuccessful even as the general problems they tackle seem to improve.

While institutional change is often rooted in social conditions evolving outside the criminal justice system, nevertheless, specific constructive changes have proximate causes and, within bounds, can be shaped. They result from the work of individuals, which translates vague sentiments into specific agendas. Thus Caleb Foote, Herbert Sturz, and Daniel Freed should properly be honored for having mobilized bail reform in the 1960s, and if many of their programs have not developed as hoped, the issues they raised remain permanent concerns of criminal justice officials and the general effort has yielded successes. Similarly, Senator Sam Ervin can be credited with flushing the issue of delay from its hiding place in judicial chambers and, in so doing, with significantly altering the ways the problem of delay and the management practices of the courts are considered.

A Problem-Oriented Approach

In the case studies, we found that formal policies often masked differences that those who administer the law inevitably consider, and that a preoccupation with formal policy ignored factors that affect implementation. As a consequence, new policies often missed their mark. One partial answer to this dilemma is a problem-oriented approach.[1] This approach embraces a concept of responsive law and fosters a consumer perspective on the courts and in so doing identifies problems as perceived and actually experienced by those who daily use and work in the courts. It insists upon a realism and a sensitivity to the details of administration. As such, it can focus attention on solutions to concrete problems. For instance, much of the current drive for sentence reform stems from a concern with the disparity that *might* occur under laws that permit considerable judicial discretion. Yet, in fact, in many states the source of concern stems from occasional instances of racial discrimination, obviously an intolerable problem but one that can easily be lost in the drive to reduce disparity in general. Similarly, our examination of delay indicated that it is a multifaceted issue, and that the term is used to refer to a number of quite different types of problems not likely to be adequately addressed by across-the-board time limits. Informal public defender policies giving priority to detained clients, prosecutorial priorities on career criminals and major offenses, judicial willingness to hear some motions on the phone, and the like may in the long run be more effective in dealing with some of the problems of delay than sweeping rules regarding time limits.

As simple as the idea of a problem-oriented approach is, policy makers prefer pursuing bold new programs to making incremental and unexciting adjustments in the administration of existing ones. One reason is simplicity; it is easier to understand the formal theory of a new policy than to master the intricacies of mundane practice and attack the practices of powerful public officials. But there is another reason. When dealing with issues of crime, public officials and financial supporters of reforms often feel a need to offer dramatic solutions to counteract what they perceive as the public's feeling of crisis and widespread disenchantment with the courts. Yet opinion surveys reveal something of the opposite.[2] While the public expresses a generalized disappointment with the courts, it nevertheless holds judges in high esteem. Furthermore, very few people have any detailed knowledge about the operations of the courts, and fewer still express clear preferences for specific policies. At first glance, one might be disappointed with such findings that suggest public apathy. Yet, they also suggest opportunity. The public perceives a problem but still has confidence in the courts. And no significant segment of the public insists that it has the answers. Here officials possess a freedom to experiment that they often do not have on other public issues.

Sources of Change

A Strategy of Rights. During the past three decades, the courts have been involved to an unprecedented extent in implementing significant changes in complex public institutions. Initiated by *Brown* v. *Board of Education,* this trend was accelerated by federal court supervision of school desegregation efforts and has since spread to other issues as well. In the criminal justice system, most obvious is the Supreme Court's ruling in *Gideon* v. *Wainwright* and subsequent decisions expanding right to counsel, an innovation that has dramatically and permanently altered the American criminal justice process.[3]

Despite—or perhaps because of—these successes, within the past few years a small library has been published that argues that the courts possess neither the competence nor the capacity to resolve social issues.[4] The argument is often buttressed by examples that show that public interest litigation has not succeeded, or that victories have unleashed consequences far worse than the original problems.[5]

While there is much truth in these criticisms, withdrawal by the courts is not warranted. One reason why many social issues come before the courts is that other institutions—legislatures, administrative agencies, private organizations—either cannot or will not address them. Many pleas are made on behalf of groups with limited access to the political process: minorities, the poor, the mentally ill, and so on. More generally, courts can play a special role in protecting individuals against powerful and impersonal institutions in a complex society.[6] Finally, solutions may be cumbersome regardless of who devises them.

Critics point to several features of public-interest cases that make them ill suited for the courts: the polycentric issues, the large numbers of people affected, the need for continuing supervision, the financial implications, and the myriad decisions required. But historically, common-law courts have entered and charted vast realms of social life, expanding law to meet new social conditions. Expansion of tort law in reaction to the industrial revolution is one example. Judicial responses to railroad reorganization, bankruptcy, and antitrust legislation are others. Judicial response to public-interest litigation in the bureaucratic welfare state is in this tradition. Judges have long supervised complex issues.[7] Indeed, the appointment of special masters, use of consent decrees, and other devices were developed by judges in an earlier era for quite different issues. And over a long period, the Court of Appeals in England has begun to fashion a law of sentencing to guide judges. Above all, critics of court-mandated changes often fail to consider that the primary function of court decisions that enunciate new standards is *general*; rules should serve as guides to issues beyond the immediate parties in the particular case.

Cornell sociologist and law professor James Jacobs has traced the strategy of litigation in the prisoners' rights movement and concluded that it has had a significant effect on prison administration. It has contributed to the marked decline in brutal and degrading practices, improved health and educational facilities, expanded religious freedom, led to fairer administration of punishments and rewards, and significantly reduced overcrowding. More generally, it has fostered recruitment of more competent administrators, made the public more aware of conditions in prisons, and made legislatures more responsive to the problems of prisons.[8]

It is important to note that decisions in prisoners' rights and other similar cases are not limited to a handful of activist judges, nor do they focus on isolated problems. They have been made by judges of all political and legal persuasions in all regions of the country. As of mid-1980, the prison systems of thirteen states and the major prisons in another dozen states were under district court supervision, and numerous other prisons were subject to a variety of court orders as well. Typically, judges in these cases view their decisions as long-overdue responses to problems involving violations of basic constitutional and legal rights. Indeed, despite the fact that often they have been the defendants, many wardens and corrections commissioners have quietly welcomed these suits, hoping that litigation would accomplish what they have been unable to do by themselves.

Turning to the issues considered in this book, we saw that the courts were often an important impetus for change. In Oakland, a federal court order seeking to remedy unconstitutionally crowded jail conditions was the proximate cause of an early move to liberalize pretrial release. In New York City numerous suits have challenged all aspects of pretrial detention and as such have served as an impetus to expand pretrial release and to expedite the handling of cases. A 1980 decision by a sharply divided California Supreme Court may have set the stage for expanded consideration of bail. In *Van Atta* v. *Scott*,[9] the Court held that the prosecutor must bear the responsibility of showing that defendants should not be released on their own recognizance. Rights litigation has had an important indirect effect on sentencing policy but as of yet has not played a major role. By circumscribing the powers of the old Adult Authority, the California Supreme Court served as a catalyst for legislative changes but did not itself fashion the beginnings of a new sentencing law. In response to legislative innovations in the late 1970s, the Minnesota Supreme Court is now beginning to develop appellate review of sentences. To date, the effect of court decisions with respect to speedy trial has been to defer to the legislature.

Litigation. Litigation is well suited to pursuing change in complex institutions. It is problem specific: it focuses on particular problems and individuals. It is ameliorative: remedies can be tailored to fit specific conditions and circumstances. It is incremental: it proceeds by steps, leading to concrete remedies. It is experimental: if one remedy fails, another can be substituted, or a successful approach can be used again and enlarged. And litigation is relatively inexpensive: it does not require that legislative majorities be mobilized. In contrast, legislation is general. In order to pass, it must often be left vague. It is time-consuming and costly. And because legislatures are keenly aware of public opinion, it tends to elevate the atypical—the newpaper headline—to the usual.[10] Finally, legislative failures are more difficult to acknowledge, let alone change, than court orders.

Like all other strategies, litigation is not a recipe for success. Indeed, in a great many areas it has perhaps exacerbated racial problems. Still, litigation is especially suited to pursuing changes in the *legal* process. Here the courts are on their home territory. And the issues of pretrial release and speedy trial both involve specific constitutional guarantees. Expanded interpretation of these provisions could hardly be regarded as judicial usurpation. Similarly, while some sentence decisions have been

Toward a Strategy for Change

based on interpretations of the constitutional ban on cruel and unusual punishment, the opportunity for expanded court action in this area is more appropriately under the equal-protection and due-process clauses and in conjunction with legislative guidelines. Indeed, restrictions on the power of public officials are the acknowledged aim of the Bill of Rights and the Fourteenth Amendment. Litigation as a strategy for *criminal justice reform* builds upon an especially strong constitutional foundation and legal tradition.

Adversarial combat asserting rights is the hallmark of the judicial process. While appellate review of sentences would open up a whole new area for the courts, the fashioning of a common law of sentences is similar to a great many issues that courts have dealt with in the past and is one with which the generally timid English courts have had modest success.

The courts are equipped to deal with rights litigation in the classic problem-reduction manner, moving first to tackle the most obvious and egregious problems, and from there evolving general doctrines. Indeed, the Supreme Court has already outlined the detailed steps trial courts must take in administering the death penalty, and at the other end of the severity continuum, appellate courts have upheld administrativelike practices in traffic cases. We might reasonably expect the courts to turn their attention to the great number of problems that fall between.

Court action in the areas discussed here does not preclude legislative action and should in fact stimulate it. I have emphasized the potential importance of the courts in an effort to counter the growing belief that courts do not have the capacity to tackle complex social problems. New bureaucratic institutions that try to prop up shortcomings of the criminal court system (for example, diversion programs, pretrial release agencies) are not compatible with the adversary process. Slow, incremental adjustments aimed at concrete problems are preferable to wholesale changes based upon the illusions of formalism. Finally, the solution to the failings of the adversary process is strengthening the process, not trying to circumvent it.

Of course, court decisions, like other formal pronouncements, can have very little effect. There is no assurance that litigation will yield immediate results; still, court decisions have the potential for producing significant indirect effects, and these may have the greatest impact. A court decision creating or expanding a right provides an important vehicle for dramatizing an objective and for shaping expectations.

Litigation supported by the American Civil Liberties Union under the direction of Alvin Bronstein has yielded significant direct effects. But perhaps more important is its indirect impact. Litigation and the threat of litigation can force recalcitrant legislators to consider problems they readily acknowledge but prefer to ignore. It provides corrections officials with new arguments for more resources. It stimulates thought about alternative forms of sentencing. Litigation should be viewed as a means for political change, not as an alternative to it. Although there are problems inherent in such a strategy,[11] the history of the indirect role litigation has played is impressive.[12]

Although many social action organizations (for example, NAACP, Jehovah's Witnesses, American Civil Liberties Union, Nader's Raiders, American Jewish Congress) have pursued litigation effectively, public defender organizations have not systematically used it. Founded as volunteer groups to represent indigent defendants, they are now often publicly supported and staffed by full-time, salaried attorneys. Despite this growth, they remain limited to their most obvious and immediate task, defending individual clients. Rarely have PD offices represented the collective interests of their clientele. Several factors account for this. First, a sense of professionalism prevents public defenders from "using" clients to test issues. Second, their resources are limited. Third, PD organizations are extremely vulnerable to political pressure. Fearing reprisals against their clients in ways that would be impossible for them to document (for example, harsh sentencing, high bond, prosecutorial refusal to bargain, being placed last on the court calendar), public defenders are reluctant to file class-action suits or pursue public-interest litigation.

Perhaps spurred by the reform efforts of their civil counterparts in the federally sponsored legal services offices, some PD offices are beginning to change. These exceptions suggest the *potential* of an organized and aggressive public defender movement. In recent years, suits brought by the public defense organization in New York City, the Legal Aid Society, have precipitated a number of important changes in release procedures and jail administration and have increased the ability of defense attorneys to meet with their detained clients. Although there have been negative reactions (the society's budget has been threatened, and it has been frequently criticized by judges for grandstanding), the society has weathered this criticism, in part because of the solidarity of its membership (its unionized attorneys have upon occasion gone on strike for improved working conditions), and in part because of the strong backing of the prestigious Association of the Bar of the City of New York and leading establishment law firms.

In the 1970s, the Los Angeles County Defender Organization initiated a series of actions to improve the status of indigent defendants as a group. These suits attacked policies concerning attorney-client communication in the county jail, pretrial detention of juveniles, and jury selection. They caused a great deal of furor, including the resignation of the head of the defender organization and the threat of a cutback of funds by the County Board of Supervisors, and won mixed victories in the courts. But according to one scholar who watched this process, "a public defender agency has the capacity to [successfully] initiate, and to sustain, litigation in support of desired policy goals."[13]

Such activity by public defender organizations remains the exception. But if the concerns of the National Legal Aid and Defender Association (NLADA) are any indication of the future, we can expect to see increased use of litigation in policy formation, increased lobbying, and work within legal associations, commissions, and the like, to draft model codes, procedures, and standards.

But PDs must resist the backlash and pressure that will inevitably follow their entry into the policy-making arena. One strategy is to organize the effort on a national level, involving local offices only when necessary. Another is to further insulate local offices from reprisal. Increasingly, and particularly in the larger cities, defender organizations are several stages removed from the electoral process, and most national legal organizations, commissions, and prominent jurists support this trend. PDs will be able to pursue a strategy of policy change through litigation only if they are well funded, but public funding is and will probably continue to be earmarked for representatives of individual clients. Thus, if PDs are to be successful, they will have to follow the strategy of the NAACP Legal Defense Fund, prisoners' rights groups, and other organizations, and obtain funds from private sources.

R and D for the Courts. The Vera Institute in New York City, considered at some length in this book, is something of a research and development arm of the criminal justice system in that city. Initially founded to administer a privately funded experimental pretrial release program, Vera has branched out to tackle a number of problems in New York's courts.

While commitment to trial and error is a cliché, in fact few

218

organizations have the capacity and integrity to admit error and try again. Yet the Vera Institute has done just this. It painstakingly developed a new approach to pretrial release, and once the program achieved success, it transferred its administration to a city agency. But, unlike so many initiators of change, Vera remained interested, and when its evaluations revealed that the project was faltering, it regained control from the city. Even now, years later, it continues to monitor informally the new pretrial release agency. Similarly, Vera pioneered in promoting pretrial diversion. Here, too, after a period of incubation, it helped establish an independent program and then continued to monitor its activities. When problems appeared, Vera helped the program rethink its mission, withdraw from the diversion field entirely, and redirect the work of its staff. Vera's work has also involved it in hundreds of other reform efforts, most small and unnewsworthy in themselves, but when viewed in the aggregate, substantial.

Vera's success is due to a number of factors. Foremost is the integrity and vision of its founder and long-time executive director, Herbert Sturz. Under his leadership, the Vera staff developed an immense self-confidence that has allowed it to maintain a truly experimental outlook, admit shortcomings, and rethink solutions. As impressive as he is and as important as he became, Sturz always administered an open organization, recruiting bright young people and propelling them into positions of authority if their work warranted. This openness has facilitated a continuous stream of new ideas and healthy self-criticism. By now this mode of operation has become institutionalized, and Vera has continued to prosper and exert perhaps even greater influence under the leadership of Sturz's successor, Michael Smith. In this successful transfer of leadership, Vera's influence has become institutionalized. The more significant innovation is the innovation-producing institution itself.

Given Vera's success in New York City, and its well-deserved national reputation, it is both surprising and disappointing that similar organizations have not emerged elsewhere. While there are some close comparisons, none to my knowledge has developed into the powerful and continuing force for change that Vera has. Some, such as the Institute for Social Research and Law (INSLAW) in Washington, D. C., provide important research studies assessing their city's criminal justice system but have not initiated new programs suggested by this research. Others initiate new programs, but lack the capacity to continue to nurture them carefully. Still other national citizen-based organizations mobilize their local chapters around particular issues, but these efforts fade into quiescence once the immediate concern has passed.

Perhaps one reason why the Vera idea has not been replicated is because of the establishment of LEAA in 1968 and the creation of state planning and regional planning units. In many respects, the mission of these SPAs and RPUs is similar to Vera's self-appointed task. There are, however, important differences. While Vera is a private institution outside the formal structure of the courts, these LEAA-spawned agencies are under the direct authority of boards representing the very criminal justice agencies they are charged with changing. And while Vera has remained independent of local criminal justice agencies, and at times has had a voice co-equal to them, as a rule the SPAs and RPUs have been captured by these agencies.[14]

From the beginning, LEAA critics maintained that these state and local units had been created only in order to obtain federal funds and were not serious planning agencies or strategically situated to assume leadership. The aftermath of the demise of LEAA suggests that these critics were correct. Most states have drastically cut back or abolished the units altogether, even though crime policy remains high on the public's agenda. Efforts by those few private foundations interested in criminal justice have not fared much better. They too have tended to support programs that are not well integrated into the community, and they rarely address the issue of institutionalizing new programs, so that, once the extraordinary funding ceases, their programs will remain as continuing sources of innovation. Perhaps the death of LEAA will stimulate greater concern over ways to institutionalize the R and D function for courts. If so, the Vera Institute—its very existence more than any particular program it has developed—stands as a useful model to reflect on.

A Modest National Effort. Despite the problems of LEAA, there may be a role for a modest national agency concerned with innovation in the criminal justice system, which might support limited experimental efforts in a few jurisdictions. Such an effort might be more successful than trying to encourage all the states to develop systemwide planning and hoping that experimental programs will emerge. This idea is by no means novel. Under LEAA's discretionary grant program, a portion of action grant funds was directly awarded by the national office for experimental programs of its own selection, rather than passed to the states on a formula basis for their distribution. Certainly, appropriating large sums of money to the states, which claim it more or less as a matter of right, is not an obvious way to try to stimulate innovation. Any national policy in this area should be designed to maximize true experimentation and evaluation. This is most likely to be achieved through the support of a few well-chosen state and local agencies. One approach might be to stimulate the growth of Vera-like organizations, which in turn could serve as continuing sources of R and D in their own local criminal justice systems.

Symbols and Courts

Throughout this book we have seen that many policies offer symbolic rather than substantive change. But it is important not to denigrate the importance of symbols altogether. Symbols can be potent instruments in fostering the political ends of both quiescence and change. At one and the same time they can give the appearance of meaningful policy without the substance and can serve as instruments of political mobilization, flags around which to rally. Federal court decisions served both these functions in the American South in the 1950s and 1960s. The decisions gave the appearance of desegregation, but there was little change. On the other hand, they served as symbols around which political action was mobilized, and this action did have an effect.

Formal policies exist on two levels simultaneously. They are promoted as solutions even though they do little to alter conditions, and they mobilize support for change. Thus, as Stuart Scheingold has so eloquently argued, a quest for rights is most likely to be effective as part of a political effort, not as a substitute for politics.[15]

Symbols also shape aspirations. It is a legitimate aim of government to express the aspirations of the community, and law—especially the criminal law—is a potent force in the pursuit of this aim. Laws, court rituals, and criminal procedure are all designed in part to promote morality and education, to strengthen a sense of community, and to assert authority. But there is a constant danger of overextension, of making false promises, exaggerating expectations, and intruding too far into private affairs. The danger of the corruption of this symbolic expression of aspirations is greatest when policies are aimed at seemingly intractable problems, such as crime. Much of criminal law scholarship concerns itself with revealing overextension, arrogance, and abuse in the criminal law. Even as we recognize these dangers, we must also recognize that the criminal justice system will pursue problems beyond its capacity to solve, and that however carefully and conscientiously the courts act, they will appear to fail. The reason is that they are charged with coping with one of society's intractable problems even as they do not have the capacity to alter significantly the conditions that give rise to this problem. Still, within their limited sphere of influence, what they do and how they appear to act is important.

I argued earlier that plea bargaining may not be the major problem so many think it is, that problems of sentence disparity may not be so extensive and significant as is commonly thought, and that a host of other problems turn out, when placed in perspective, to be something other than what they appear. But appearances count for a great deal. This obvious fact is often neglected by both court officials and reformers.

Most courtroom practices take place in such a manner as to almost ensure the appearance of injustice. Shabby facilities, incomprehensible language, uncommunicative officials, and lack of decorum are the rule in courts. This problem is a concentrated microcosm of an inegalitarian society that professes egalitarian ideals. Compare, for instance, the nature and demeanor of proceedings in federal courts, which deal primarily with civil litigation involving large sums and middle-class litigants, and those state courts whose primary work is criminal cases, whose clientele is poor. Still, even within the constraints imposed by social structure, there is room for significant improvement.

For example, nowhere is the disjuncture between reality and the appearance of injustice in the court so obvious as with regard to race. As with so many other features of the courts, here, too, problems are more complicated than they appear. Racial slurs and patronizing language are heard frequently in the courtrooms, and the impression of racial discrimination persists despite the enormous progress that has been made.

Increasingly, the accused, victims, and witnesses are black or brown or red, while authority remains white. While the courts are not responsible for this imbalance, nevertheless, one of their functions is to serve as a symbol of equality and community. This function is seriously eroded by the impression of racial discrimination. Again, appearances count. Even if the problem is not of the courts' own making, the means of amelioration are within their grasp.

Conclusion

In recent years, it has become commonplace to compare the criminal justice system to a hydraulic process that will inevitably restore original equilibrium in reaction to all efforts to make changes; in short, to assert that nothing works. While this book has shown that courts are interdependent and adaptive, it has also shown that changes can and do take place. Efforts to institute change are not invariably offset by corresponding adjustments. This last chapter has identified several change strategies that suggest themselves as particularly effective. While they are not recipes for success, neither are they pipe dreams. They are approaches that have yielded substantial effects and are applicable to a variety of problems. If agents of change pierce the veil of formalism, adopt lower profiles, adjust their expectations, and move to take ameliorative actions, they might have more effect than is imagined. There are no easy answers. But if we are fortunate, we might make some halting and partial advances.

Notes

Part 1 Introduction

Chapter 1: The Courts and Change

1. Jerome Frank, *Courts on Trial* (New York: Atheneum, 1963). See especially chapter 6, "The 'Fight' Theory Versus the 'Truth' Theory." For similar and more recent indictments, see Marvin Frankel, *Partisan Justice* (New York: Hill and Wang, 1980), and Lloyd L. Weinreb, *Denial of Justice* (New York: Free Press, 1977).

2. Richard A. Posner, *Economic Analysis of Law* (Boston: Little, Brown, 1973), p. 321.

3. For a thoughtful and critical assessment of the dominant judicial role in the United States, see Abraham Goldstein, *The Passive Judiciary* (Baton Rouge, La.: Louisiana State University Press, 1981).

4. Abraham Goldstein, "The State and the Accused: Balance of Advantage in Criminal Procedure," *Yale Law Journal*, 69 (June 1960): 1149–99.

5. For an interesting comparison of American and Continental criminal procedure, see Weinreb, *Denial*.

6. See Judge Marvin Frankel's discussion of the limited authority of courts to review criminal sentences in *Criminal Sentences* (New York: Hill and Wang, 1973).

7. Ibid., p. 1.

8. Karl Menninger, *The Crime of Punishment* (New York: Viking, 1968).

9. James Q. Wilson, *Thinking About Crime* (New York: Basic Books, 1975).

10. John Hogarth, *Sentencing as a Human Process* (Toronto: University of Toronto Press, 1971).

11. Herbert L. Packer, *The Limits of the Criminal Sanction* (Stanford: Stanford University Press, 1968).

12. Martin A. Levin, *Urban Politics and the Criminal Courts* (Chicago: University of Chicago Press, 1977).

13. Malcolm M. Feeley, *The Process Is the Punishment: Handling Cases in a Lower Criminal Court* (New York: Russell Sage Foundation, 1979).

14. James Eisenstein and Herbert Jacob, *Felony Justice: An Organizational Analysis of Criminal Courts* (Boston: Little, Brown, 1977). See also James Q. Wilson, *Varieties of Police Behavior* (Cambridge, Mass.: Harvard University Press, 1968), pp. 227–76; and David Bayley, ed., *Police and Society* (Beverly Hills: Sage Publications, 1977).

15. Jay Wishingrad, "The Plea Bargain in Historical Perspective," *Buffalo Law Review*, 23 (winter 1974): 499–527.

16. Julius Goebel and T. Raymond Naughton, *Law Enforcement in Colonial New*

York: A Study in Criminal Procedure (1664–1776) (New York: The Commonwealth Fund, 1944).

17. Albert W. Alschuler, "Plea Bargaining and Its History," *Law and Society Review,* 13 (winter 1979): 211–45.

18. Raymond Moley, *Politics and Criminal Prosecution* (New York: Minton Balch, 1929), pp. 160–61.

19. Ibid., pp. 58–65.

20. Milton Heumann, "A Note on Plea Bargaining and Case Pressure," *Law and Society Review,* 9 (spring 1975): 515–27; and Malcolm M. Feeley, *Process,* pp. 244–77.

21. Lawrence M. Friedman, "Plea Bargaining in Historical Perspective," *Law and Society Review,* 13 (Winter 1979): 247–59. For a review of some of these earlier accounts of the criminal process and a useful list of original sources, see Peter F. Nardulli, *The Courtroom Elite* (Cambridge, Mass: Ballinger, 1978); and Mark Haller, "Urban Crime and Criminal Justice: The Chicago Case," *Journal of American History* 57 (1970): 619–35.

22. Malcolm M. Feeley, "Plea Bargaining, Professionalism and Progress" (unpublished manuscript, University of Wisconsin, 1981).

23. Friedman, "Plea Bargaining," p. 257. See also Lawrence M. Friedman and Robert C. Percival, *Roots of Justice* (Chapel Hill: University of North Carolina Press, 1981).

24. Samuel Dash, "Cracks in the Foundation of Criminal Justice," *Illinois Law Review,* 66 (July-August 1951): 385–406, p. 392.

25. *Gideon* v. *Wainwright,* 372 U.S. 335 (1963).

26. Feeley, *Process,* pp. 244–77.

27. Quoted in ibid., *Process,* p. 266.

28. Anthony Japha, *The Effects of the 1973 Drug Laws on the New York State Courts: A Staff Report of the Drug Law Evaluation Project* (New York: The Association of the Bar of the City of New York and the Drug Abuse Council, August 1976), pp. 7–8.

29. Jack Hausner and Michael Seidel, *An Analysis of Case Processing Time in the District of Columbia Superior Court* (Washington, D.C.: INSLAW, Inc., 1981), no. 15, p. 51.

30. Geoff Gallas and Michael Lampasi, "A Code of Ethics for Judicial Administration," *Judicature* 61 (February 1978): 311–17, p. 312.

31. Ramsey Clark, *Crime in America* (New York: Pocket Books, 1970), pp. 177–78.

32. Commissioner Archibald Murray, quoted in Feeley, *Process,* p. 266.

33. Thomas Church, Alan Carlson, Jo-Lynne Lee, and Teresa Tan, *Justice Delayed: The Pace of Litigation in Urban Trial Courts* (Williamsburg, Va.: National Center for State Courts, 1978). See also David Neubauer, Marcia J. Lipezt, Mary Lee Luskin, and John Paul Ryan, *Managing the Pace of Justice* (Washington, D.C.: National Institute of Justice, September 1981).

34. *Felony Arrests: Their Prosecution and Disposition in New York City's Courts* (New York: Vera Institute and Longman, 1981), p. 134.

35. Ibid.

36. Ibid,., p. 136.

37. Ibid., p. xxi.

38. Alfred Blumstein, Jacqueline Cohen, and Daniel Nagin, "Report of the Panel," in *Deterrence and Incapacitation: Estimating the Effects of Criminal Sanctions on Crime Rates* (Washington, D.C.: National Academy of Sciences, 1978), pp. 46–47.

Notes

39.   Blumstein et al., "Report," pp. 78–79. See also Jacqueline Cohen, "The Incapacitative Effect of Imprisonment: A Critical Review of the Literature," in Blumstein et al., *Deterrence*, pp. 187–243.

40.   For an enlightening analysis of crime, corruption, and the urban criminal courts at the turn of the century, see Mark Haller, "Urban Crime and Criminal Justice," pp. 619–35. More generally, see Harold Gosnell, *Machine Politics: Chicago Model*, 2nd ed. (Chicago: University of Chicago Press, 1968); Robert M. Fogelson, *Big-City Police* (Cambridge: Harvard University Press, 1977); John A. Gardner and David J. Olson, eds., *Theft of the City* (Bloomington: Indiana University Press, 1974).

41.   Robert Hermann, Eric Single, and John Boston, *Counsel for the Poor: Criminal Defense in Urban America* (Lexington, Mass.: Lexington Books, 1977), p. 153.

42.   For discussions of the incapacitative effects of imprisonment, see David F. Greenberg, "The Incapacitative Effect of Imprisonment: Some Estimates," *Law and Society Review* 9 (Summer 1974); reports on the continuing research of the Rand Corporation, Joan Petersilia, Peter W. Greenwood, and Marvin Lavin, *Criminal Careers of Habitual Felons* (Washington, D.C.: National Institute of Law Enforcement and Criminal Justice, July 1978); and Peter Greenwood, *Selective Incapacitation* (Santa Monica, Ca.: Rand Corp., November 1981). See also Mark H. Moore, *Buy and Bust* (Lexington, Mass.: Lexington Books, 1977).

43.   See, for example, Blumstein et al., "Report," pp. 19–80; and Report by the Comptroller General of the United States, *Impact of the Exclusionary Rule on Federal Criminal Prosecutions* (Washington, D.C.: General Accounting Office, April 19, 1979). More generally, see the symposium on the exclusionary rule in *Judicature* 62, nos. 2, 5, 7, 9 (1979).

Part 2   Some Innovative Programs

Introduction: The Process of Planned Change

1.   In the following discussion and subsequent case studies I have been influenced by the perspective developed by Jerald Hage and Michael Aiken in *Social Change in Complex Organizations* (New York: Random House, 1970).

2.   Ibid., pp. 32–49.

3.   Ibid., pp. 93–106.

Chapter 2: Bail Reform

1.   Quoted in Ronald L. Goldfarb, *Ransom: A Critique of the American Bail System* (New York: Harper & Row, 1965), p. 24. From Sir Frederick Pollock and Frederic William Maitland, *A History of English Law*, 2nd ed. (Cambridge: The University Press, 1898) 2: 58–90.

2.   Arthur L. Beeley, *The Bail System in Chicago* (Chicago: University of Chicago Press, 1966, first published 1927).

3.   Wayne H. Thomas, Jr., *Bail Reform in America* (Berkeley: University of California Press, 1976), pp. 42, 74. A national survey by the American Bar Association, reported by Lee Silverstein, found detention rates of 51, 51, and 46 percent in large,

medium, and small counties, with an overall average of about 50 percent. See Lee A. Silverstein, "Bail in the State Courts: A Field Study and Report," *Minnesota Law Review* 100 (1966): 621–52. These figures comport with other figures for 1962 reported by Thomas, pp. 37–42.

4. Even assets cannot always attract a bondsman. Ronald Goldfarb reports that, during the civil rights demonstrations in the South in the 1960s, large numbers of arrestees with the means to pay a bondsman were nevertheless unable to secure release—because bondsmen refused to do business with them. Elsewhere, if the police do not want a particular arrestee released, bondsmen are usually obliging. See Malcolm M. Feeley, *The Process Is the Punishment* (New York: Russell Sage Foundation, 1979). Also, some stay in jail because bondsmen do not want to be bothered with low bail cases.

5. Quoted in *An Evaluation of Policy Related Research on the Effectiveness of Pretrial Release Programs* (Denver: National Center for State Courts, 1975), pp. 16–17, from Roscoe Pound and Felix Frankfurter, eds., *Criminal Justice in Cleveland: Reports of the Cleveland Foundation Survey of the Administration of Criminal Justice in Cleveland, Ohio* (Cleveland: Cleveland Foundation, 1922); repr. (Montclair, N.J.: Patterson Smith, 1968), pp. 290–92.

6. Beeley, *Bail System*, p. x.

7. Wickersham Commission, *U.S. National Commission of Law Observance and Enforcement* (Washington, D.C.: GPO, 1931).

8. *Gideon* v. *Wainwright*, 372 U.S. 335 (1963) extended the right of free counsel to those accused of felonies; *Argersinger* v. *Hamlin*, 407 U.S. 25 (1972)—right of free counsel in all criminal proceedings where the prospect of confinement exists; *Griffin* v. *People of the State of Illinois*, 351 U.S. 12 (1956)—access to free transcripts in criminal appeals; *Douglas* v. *People of the State of California*, 372 U.S. 353 (1963)—right to counsel in criminal appeals. The significance of the famous school desegregation case, *Brown* v. *Board of Education*, 347 U.S. 483 (1954) in elevating the problem of racial discrimination to a national policy issue is well known.

9. Caleb Foote, "Compelling Appearance in Court: Administration of Bail in Philadelphia," *University of Pennsylvania Law Review* 102 (1954): 1031–79. See also Caleb Foote, "The Coming Constitutional Crisis in Bail," Ibid. 113 (May-June, 1965): 959–1185. See also Anne Rankin, "The Effect of Pretrial Detention," *New York University Law Review* 39 (June 1964): 631–36; Eric Single's appendix in *Plaintiff's Memorandum in Johnny Roballo, et al.* v. *The Judges and Justices of New York City Criminal Court, et al.*, 74 Civ. 2113-MEL (1974); Feeley, *Process*.

10. *Programs in Criminal Justice Reform, Ten-Year Report, 1961–1971* (New York: Vera Institute of Justice, 1972), p. 31.

11. Figures through 1973 are reported by Lee Friedman in "The Evolution of Bail Reform," *Policy Sciences* 7 (1976): 281–313. Estimates on the number of programs as of 1980 were supplied by the National Pretrial Resource Center, Washington, D.C.

12. Malcolm M. Feeley, "The New Haven Redirection Center," in Richard Nelson and Douglas Yates, eds., *Innovation and Implementation in Public Organization* (Lexington, Mass.: Lexington Books/D.C. Heath, 1978), pp. 39–67. See also Paul Lazarsfeld, *An Evaluation of the Pretrial Services Agency of the Vera Institute of Justice* (New York: Vera Institute of Justice, December 1974).

13. Floyd Feeney, *The Police and Pretrial Release* (Lexington, Mass.: D.C. Heath, 1982).

14. Daniel J. Freed and Patricia M. Wald, *Bail in the United States* (Washington, D.C.: Department of Justice and Vera Foundation, 1964).

Notes

15. Thomas, *Bail Reform*. For instance, in 1962 in Philadelphia no one accused of a felony was released on his or her own recognizance, but by 1972 fully one-third of accused felons were so released. He reports that other cities experienced changes of similar magnitude. See pp. 40–41, 70–71. This trend has continued.

16. Nan C. Bases and William F. McDonald, *Preventive Detention in the District of Columbia: The First Ten Months* (Washington, D.C.: Georgetown Institute of Criminal Law and Procedure and Vera Institute of Justice, 1972). In 1981 the Supreme Court affirmed without comment a lower-court decision upholding the constitutionality of this provision. U.S. (1981).

17. Charles E. Ares, Anne Rankin, and Herbert Sturz, "The Manhattan Bail Project: An Interim Report on the Use of Pre-Trial Parole," *New York University Law Review* 38 (1963): 67–95. See also Friedman, "Evolution," pp. 281–313, 290–91.

18. Lee Friedman, in "Evolution," has presented similar criticisms.

19. Ibid., p. 296.

20. Ibid., p. 298.

21. Ibid.

22. Pretrial Services Agency: Vera Institute of Justice, *A Report on the Operation of the Pretrial Services Agency During the Period Between June 1974 and November 1975* (New York: Vera Institute of Justice, n.d.), p. 2.

23. Ibid., p. 3.

24. *When Should a Release Agency Intervene? Analysis of a Pilot Program of Making ROR Recommendations Immediately after Arraignment* (New York: NYC Criminal Justice Agency, 1977), p. 3.

25. Ibid., p. 17.

26. Lazarsfeld, *Evaluation*.

27. Ibid., pp. 112–13.

28. These figures are slightly lower than those reported in New York City in the late 1960s and early 1970s but are slightly higher than those reported for some of the early postrelease notification efforts. See *Further Work in Criminal Justice Reform: A Five-Year Report from the Vera Institute of Justice, 1971–1976* (New York: Vera Institute of Justice, 1977), pp. 20–21. Precise figures on notification are not given. These figures are the same as the best-informed estimates of the national average FTA rate of 9 percent and considerably lower than the average rate in the nation's second largest city, Chicago, estimated to have a 17 percent FTA rate. See *An Evaluation of Policy Related Research*, p. 72; Thomas, *Bail Reform*, pp. 98–109.

29. See Feeley, *Process*, pp. 201–15.

30. In 1975, the Los Angeles Superior Court's OR Unit had a budget of $775,000, while in 1975 the annual cost of the Brooklyn operation of the release component of the city's Pretrial Services Agency was $1,450,044. Los Angeles has an elaborately decentralized system of courts, which means that resources must be spread thin. Thus, the difference in cost is even more pronounced than the figures suggest. Pretrial Services Agency, *Report*, p. 90; *O.R. Statistics End of Year Report*, annual report for 1975 (Los Angeles: Superior Court Own Recognizance Division, 1976). Obtained from John Toomey, head of the Los Angeles OR Program.

31. Author interview with Gene Babb, director of the San Francisco OR Unit, January 1978.

32. D. L. Kuykendall and R. W. Deming, "Pretrial Release in Oakland, California," mimeographed (Oakland, Calif.: Oakland Pretrial Release Project, 1967), p. 69.

33. Forrest Dill, "Bail and Bail Reform: A Sociological Study" (Ph.D. diss., University of California, Berkeley, 1972), pp. 168–88; Thomas, *Bail Reform*, p. 30.

34. Kuykendall and Deming, "Pretrial Release," p. 76.

35. See, for example, Emmet Burke, *An Evaluation of the Comprehensive Pretrial Services Delivery System in Alameda County* (Oakland, Calif.: Office of Criminal Justice Planning, 1976).

36. Ibid., pp. 88–90.

37. Ibid., p. 88. It should be noted that these FTA figures are computed in a peculiar way and can only be considered relative indicators of FTA rates.

38. Dill, "Bail," pp. 168–88.

39. Mary A. Toborg and Nathan I. Silver, *Pre-trial Release: Delivery System Analysis of Baltimore City, Maryland* (Washington, D.C.: Lazar Institute, July 1978), p. 48.

40. Thomas, *Bail Reform*, pp. 40–41.

41. Ibid., p. 165.

42. Additional evidence questioning the efficacy of special pretrial release agencies is also found in the release figures of some federal district courts. For instance, long before bail became a major concern, the federal court in Connecticut was releasing 65 percent of defendants on their own recognizance, a rate that exceeds most jurisdictions even today. Ibid., p. 162.

43. Author interview with Daniel J. Freed, March 1977.

44. Roy B. Flemming, C. W. Kohfeld, and Thomas M. Uhlman, "The Limits of Bail Reform: A Quasi-experimental Analysis," *Law and Society Review* 14 (1980): 947–76.

45. Ibid., p. 957.

46. Roy B. Flemming, *Allocating Freedom and Punishment: Pretrial Release Policies in Detroit and Baltimore* (New York: Longman, 1982).

47. Isaac Balbus, *The Dialectics of Legal Repression* (New York: Russell Sage Foundation, 1973).

48. Mary A. Toborg and Martin D. Sorin, "Pretrial Release: A National Evaluation of Practices and Outcomes," mimeographed (Washington, D.C.: Lazar Institute, August 1981).

49. Jeffrey R. Chanin, "Judicial Practice and Decision-Making in Pretrial Release," mimeographed (Providence, R.I.: Office of the Public Defender, 1977).

50. Thomas, *Bail Reform*, p. 41.

51. Lewis R. Katz, *Justice Is the Crime* (Cleveland, Ohio: Case Western Reserve University Press, 1972), p. 226.

52. Information on Milwaukee release rates was obtained from the Wisconsin Center for Public Policy, Madison, Wisconsin. I am grateful for its aid. See also Toborg and Sorin, "Pretrial Release," pp. 43–47.

53. Roy B. Flemming, "Freedom, Equity and Bail Reform: The Severity and Distribution of Pretrial Sanctions in Three Cities" (Paper delivered at the 1979 Annual Meeting of the Law & Society Association, San Francisco, Calif., May 10–12, 1979). See also Flemming, *Allocating Freedom and Punishment.*

54. These statistical objections are not easily overcome, since officials resist full-scale experiments involving random assignment. There have, however, been some imaginative efforts by several sets of ingenious researchers. For instance, Michael R. Gottfredson of the State University of New York found that widely accepted release conditions, if applied, lead to the release of people who would subsequently fail to appear, and to the detention of people who, if released, would appear. See Michael R. Gottfredson, "An Empirical Analysis of Pretrial Release Decision," *Journal of Criminal Justice* 2 (1974): 287–301. For other excellent discussions, see John S. Goldkamp, *Two Classes of Accused: A Study of Bail and Detention in America* (Cambridge:

Notes

Ballinger, 1979); Toborg and Sorin, "Pretrial Release," pp. 24–34; and Flemming, *Allocating Freedom and Punishment.*

Paul Lazarsfeld found that just one simple factor alone—whether or not the defendant could produce a telephone number at which he could be reached—was as good as any multifactor indicator. But even this indicator, he realized, was not reliable, and in conclusion he recommended that rather than trying to develop a more accurate prediction index, the agency's follow-up notification services be strengthened. Lazarsfeld's reasoning was simple: many arrests take place in the arrestee's place of residence, at which time the arresting officer could note the number of the telephone in the kitchen. From then on, he argued, follow-up reminders of court appearance dates would be made by phone. See Lazarsfeld, *Evaluation.*

55. Author interview with Bruce Beaudin at Yale Law School, March 10, 1976.
56. Chanin, "Judicial Practice," p. 59.
57. Dill, "Bail," pp. 159–68.
58. See Malcolm M. Feeley, "The Impact of Assessing the Evaluations of Courts" (Washington, D.C.: National Criminal Justice Reference Service, 1978). See also *An Evaluation of Policy Related Research.*
59. There have been some imaginative, second-best efforts to test the effects of these programs, often by scholars with limited funds taking advantage of "natural" experiments. See, for example, Gottfredson, "Empirical Analysis," and John A. Goldkamp, "Two Classes."
60. See *An Evaluation of Policy Related Research;* and Thomas, *Bail Reform.*
61. Toborg and Sorin, "Pretrial Release."
62. See *Pretrial Release: Delivery System Analysis of Baltimore County, Maryland* (Washington, D.C.: Lazar Institute, 1978), pp. 48, 53; Toborg and Silver, *Pretrial Release,* p. 47; *Pretrial Release. Delivery System Analysis of Santa Cruz County, California* (Washington, D.C.: Lazar Institute, August 1978), pp. 19–24.
63. Feeney, "The Police."
64. Project, "Preventive Detention: A Step Backward for Criminal Justice," *Harvard Civil Rights Civil Liberties Law Review,* 6 (1971): 291.
65. Toborg and Sorin, Pretrial Release," pp. 19–23.
66. See Alexander Bickel, *The Least Dangerous Branch* (Indianapolis: Bobbs-Merrill, 1962).
67. *Stack* v. *Boyle,* 342 U.S. 1, 4–5 (1951).
68. A comparative study of three quite different jurisdictions revealed that defendants generally "lacked a legal representative to act as their advocate when initial pretrial release conditions were established." Flemming, *Allocating Freedom and Punishment,* p. 8.
69. A number of recent studies have shown that for a substantial number of arrestees, the most important decisions and the harshest sanctions are extracted in connection with the pretrial release decision, and not sentencing. See, for example, Feeley, *Process;* Herbert Jacob and James Eisenstein, *Felony Justice* (Boston: Little, Brown, 1977); Flemming, *Allocating Freedom,* and Goldkamp, "Two Classes."

Chapter 3: Pretrial Diversion

1. President's Commission on Law Enforcement and Administration of Justice, *The Challenge of Crime in a Free Society* (Washington, D.C.: GPO, 1967), p. 131.
2. Elizabeth W. Vorenberg and James Vorenberg, "Early Diversion from the

Criminal Justice System: Practice in Search of a Theory," in Lloyd Ohlin, ed., *Prisoners in America* (Englewood Cliffs, N.J.: Prentice-Hall, 1973), pp. 151–66, 152.

3. See, for example, Howard Becker, *The Outsiders* (New York: Free Press, 1963), The President's Commission on Law Enforcement and Criminal Justice, *Task Force Report: The Courts* (Washington, D.C.: GPO, 1967), pp. 4–13.

4. U.S. Department of Labor, Manpower Administration, *Pretrial Intervention: Final Report*, MEL 75-02 (Washington, D.C.: GPO, 1974), p. 1.

5. Ibid., p. 2.

6. Ibid.

7. Ibid., p. 9.

8. Telephone interview with Madeleine Crohn, February 16, 1979.

9. *Court Employment Project* (New York: Court Employment Project, n.d., circa 1970–72).

10. Ibid., p. 3.

11. The Manhattan Court Employment Project (New York: Vera Institute of Justice, 1972), pp. 7–9; 52.

12. Ibid., p. 12.

13. *Pretrial Diversion from Prosecution: Descriptive Profiles of Seven Selected Programs* (New York: Vera Institute of Justice, 1978), p. 31.

14. *Pretrial Diversion*, pp. 2–3.

15. Sally Baker and Susan Saad, *Court Employment Project Evaluation: Final Report* (New York: Vera Institute of Justice, 1979).

16. Ibid., p. 276.

17. Minutes of CEP's board of trustees, December 12, 1978, as quoted in *Court Employment*, p. 284.

18. Ibid., pp. 372–73.

19. *Annual Report: Fiscal Year 1974–75* (New York: Court Employment Project, n.d.), p. 7.

20. *Pretrial Diversion*, p. 25.

21. Ibid., p. 32.

22. *Manhattan Court Employment Project*, pp. 7–8, 52.

23. For a review of this reaction to the initial positive assessment of the Court Employment Project in its 1972 Report *(The Manhattan Court Employment Project)*, see Franklin Zimring, "Measuring the Impact of Pretrial Diversion from the Criminal Justice System," *University of Chicago Law Review* ICIV (1974): 224–41, 228.

24. Elizabeth Vorenberg and James Vorenberg, "Early Diversion."

25. Paul Nejelski, "Diversion: The Promise and the Danger," *Crime and Delinquency* 22 (1976): 393–410.

26. Norval Morris, *The Future of Imprisonment* (Chicago: University of Chicago Press, 1974), pp. 9–10. Zimring, "Measuring," p. 240.

27. Baker and Saad, *Court Employment*.

28. Ibid., pp. 147–214, 155, 165.

29. Ibid., pp. 188–93.

30. All figures in above discussion are drawn from *Court Employment*, pp. 254–67.

31. Gerald F. Cox, "Project Intercept: Eighteen Months' Progress Report," April 13, 1971–October 13, 1972, mimeographed (San Jose, Calif.: Project Intercept), pp. 16–17.

32. Ibid., pp. 78–81.

33. Interview with Richard Boss, director of Project Intercept, San Jose, Calif., January 15, 1978.

34. Daniel J. Freed, Edward J. DeGrazia, and Wallace Loh, "Report to the New

Notes

Haven Pretrial Services Council: The New Haven Pretrial Diversion Program" (Paper on file at Yale Law School, New Haven, Conn., 1973), p. 9.

35. Ibid., p. 67.

36. Ibid.

37. Anthony Platt, *The Child Savers* (Chicago: University of Chicago Press, 1968); Ellen Ryerson, *The Best Laid Plans* (New York: Hill and Wang, 1978).

38. See, for example, Roberta Rovner-Pieczenik, *Pretrial Intervention Strategies* (Lexington, Mass.: Lexington Books, 1976); Joan Mullen, principal investigator, *Pretrial Services: An Evaluation of Policy Related Research* (Cambridge, Mass.: Abt Associates, December 1974); Abt Associates, *Pretrial Intervention: Final Report* (Cambridge, Mass.: Abt Associates, July 31, 1974).

39. Abt, *Final Report*, p. 21.

40. Rovner-Pieczenik, *Pretrial Intervention Strategies*, pp. 143–44.

41. Baker and Saad, *Court Employment*.

42. *Felony Arrests: The Prosecution and Disposition in New York City's Courts*, rev. ed. (New York: Vera Institute and Longman, 1981).

43. Joan Mullen and Daniel McGillis, *Neighborhood Justice Centers: An Analysis of Potential Models* (Washington, D.C.: GPO, October 1977), p. 196. See also Warren Burger, "Agenda for 2000 AD—A Need for Systematic Anticipation," *Federal Rules Decisions* 120 (1976).

44. See, for example, Joseph S. Lobenthal, *Power and Put-On* (New York: Outerbridge and Dienstfrey, 1970); Marc Galanter, "Why the 'Haves' Come Out Ahead: Speculations on the Limits of Legal Change," *Law and Society Review* 9 (1974): 63–94; Idem, "Avoidance as Dispute Processing: An Elaboration," *Law and Society Review* 9 (1974): 695–706.

45. See Roman Tomasic and Malcolm M. Feeley, eds., *Neighborhood Justice: An Assessment of an Emerging Idea* (New York: Longman, 1982).

46. See, for example, Richard Abel, "A Comparative Theory of Dispute Institutions in Society," *Law and Society Review* 8 (1974): 217–347; Richard Abel, ed., *Informal Justice* (New York: Academic Press, 1981).

47. This brief and skeptical discussion of neighborhood justice centers is not meant to diminish the seriousness of the problems that this reform addresses—making justice accessible. The problem of inaccessible justice is real, and the movement to seek ways to provide access to justice and to develop workable mediation programs as alternatives to courts quite properly commands the attention of large numbers of thoughtful scholars and practitioners throughout Western industrialized societies. See, for example, Ford Foundation, *Mediating Social Conflict* (New York: Ford Foundation, 1978); and the several volumes of the *Access to Justice* project, under the general editorship of Mauro Cappelletti, vol. I: *A World Survey* (1978); vol. II: *Promising Institutions* (pt. I, 1178; pt. II, 1979)· vol. III: *Emerging Issues and Perspectives* (1979); vol. IV: *Access to Justice in an Anthropological Perspective* (1979) (Amsterdam: Sijhoff and Noordhoff).

My hope is that reformers ponder the thoughtful discussions and suggestions that are examined in the material above.

Chapter 4: Sentence Reform

1. Cesare Beccaria, *On Crimes and Punishment* (London: J. Almon, 1767).

2. Jeremy Bentham, "Principles of Penal Law," in Bentham, *Works* (Edinburgh: W. Tait, 1837, vol. 1). More generally, see Leon Radizinowitz, *History of the English*

Criminal Law (London: Macmillan, 1948), vol. 1, and J. F. Stephen, *History of the Criminal Law in England* (London: Macmillan, 1883).

3. Douglas Hay, Albion's Fatal Tree: *Crime and Society in Eighteenth Century England* (New York: Pantheon Books, 1975), pp. 17–63.

4. Ibid. pp. 23–25; and Michael A. Rustigan, "A Reinterpretation of Criminal Law Reform in Nineteenth Century England," *Journal of Criminal Justice* 8 (1980): 205–20.

5. Harry K. Barnes and Negley K. Teeters, *New Horizons in Criminology,* 3rd ed. (Englewood Cliffs, N.J.: Prentice-Hall, 1959); David Fogel, *We Are the Living Proof: The Justice Model of Corrections* (Cincinnati: W. H. Anderson, 1975), p. 17.

6. See, for example, David J. Rothman, *The Discovery of the Asylum* (Boston: Little, Brown, 1971).

7. Edward Lindsey, "Historical Sketch of the Indeterminate Sentence and Parole System," *Journal of Criminal Law, Criminology and Police Science* 16 (1925): 9–126. See also David Rothman, *Conscience and Convenience* (Boston: Little, Brown, 1980).

8. See Michael Ignatieff, *A Just Measure of Pain: The Penitentiary in the Industrial Revolution, 1750–1850* (London: Macmillan, 1978); Michael Faucault, *Discipline and Punishment: The Birth of the Prison* (New York: Pantheon, 1978).

9. *New York Times,* January 4, 1973, sect. 1, p. 29.

10. Ibid.

11. Ibid., January 6, 1973, sect. 1, p. 1.

12. Ibid., January 10, 1973, sect. 1, p. 49.

13. Ibid., May 5, 1973, sect. 1, p. 35.

14. Ibid.

15. Ibid., January 4, 1973, sect. 1, p. 38.

16. Ibid., April 28, 1973, sect. 1, p. 1.

17. Ibid., January 11, 1973, sect. 1, p. 24.

18. Drug Law Evaluation Project, "Sentencing Patterns Under the 1973 New York State Drug Laws, Staff Memorandum" (New York: Association of the Bar of the City of New York, October 1976).

19. Ibid., p. 34; Drug Law Evaluation Project, "Crime Committed by Narcotics Users in Manhattan, Staff Memorandum" (New York: Association of the Bar of the City of New York, May 1976).

20. *New York Times,* September 8, 1973, sect. 1, p. 17.

21. Ibid., January 9, 1973, sect. 1, p. 38.

22. Ibid., May 9, 1973, sect. 1, p. 1.

23. Ibid.

24. Ibid., May 13, 1973, sect. 4, p. 5.

25. Drug Evaluation Project, "The Effects of the 1973 Drug Laws on the New York State Courts," Staff Memorandum (New York: Association of the Bar of the City of New York, August 1976), sect. 5, p. 2.

26. *People* v. *Donigan,* County Court, Suffolk Co. Ind. No. 417–74, Memorandum Opinion per Judge Signorelli, June 12, 1975.

27. Drug Law Evaluation Project, *Effects,* sect. 2, p. 4.

28. *New York Times,* January 26, 1973, sect. 1, p. 18.

29. Drug Law Evaluation Project, *Effects,* sect. 6, pp. 22–23.

30. Ibid., sect. 5, p. 2.

31. Ibid., sect. 3, p. 18.

32. *New York Times,* January 20, 1973, sect. 1, p. 28.

33. Drug Law Evaluation Project, *Effects,* sect. 2, Table 2-I, p. 2.

34. Ibid., sect. 4, p. 3.

35. Ibid., sect. 4, p. 4.

Notes

36. Philip Richardson and Bernard Gropper, "Evaluating Drug Laws and Control Policy Effectiveness: New York's Experience with the Nation's Toughest Drug Law," in Margaret Evans, ed., *Discretion and Control* (Beverly Hills, Calif.: Sage Publications, 1978).

37. Drug Law Evaluation Project, *Effects,* sect. 6, p. 12.

38. Ibid.

39. Ibid.

40. Drug Law Evaluation Project, *Sentencing,* Table XV, p. 42.

41. Ibid., p. 39.

42. Ibid., p. 54.

43. Richardson and Gropper, "Evaluating," p. 49.

44. Ibid., p. 55.

45. See, for example, *Taylor* v. *Sise,* 33 *N.D.* 2d 357 (1974), and *Carmona* v. *Ward,* cert. denied, 99 *S. Ct.* 874 (1978); Justice Marshall dissenting.

46. Petitioner's "Pre-hearing Memorandum in Support of the Application for Writs of Habeas Corpus" in *Carmona* v. *Ward* (S.D.N.Y. May 24, 1976), later reported as 436 F. Supp. 1153 (S.D.N.Y. 1977).

47. *Proceedings of the Second National Forum on Handgun Control,* Lexington, Ky.: United States Conference of Mayors, January 7–9, 1976, p. 37.

48. James A. Beha, "And Nobody Can Get You Out," *Boston University Law Review* 58 (1977): 106–208. This article represents a preliminary analysis of the effects of Bartley-Fox.

49. Speech of Judge John Fox, *Forum on Handgun Control,* pp. 31–32.

50. See Beha, "And Nobody."

51. Ibid., pp. 140–44.

52. Glen L. Pierce and William J. Bowers, *The Impact of the Bartley-Fox Gun Law on Crime in Massachusetts* (Boston: Center for Applied Social Research, Northeastern University, April 1979), p. 92.

53. Ibid., p. 91.

54. Ibid., p. 92.

55. Paul Froyd, *The Bartley-Fox Act: Impacts for Courts of the Massachusetts Gun Law,* mimeographed (Boston: Boston University Law School, April 18, 1978), p. 53.

56. Beha, "And Nobody," p. 54.

57. Ibid., p. 59.

58. Ibid., pp. 67–73.

59. David Rossman, Paul Froyd, Glen L. Pierce, John McDevitt, and William J. Bowers, *The Impact of the Mandatory Gun Law in Massachusetts* (Boston: Boston University Center for Criminal Justice, 1979), p. 8.

60. Ibid., p. 264. See also Lawrence W. Sherman, "Enforcement Workshop: The Police and the Mandatory Gun Law," *Criminal Law Bulletin* 16, no. 1: 1980: 165–66. Sherman suggests that unless legislatures appropriate funds to educate police about complex laws, inadequate enforcement is almost unavoidable.

61. Ibid., p. 8.

62. Rossman, "Impact," pp. 233–34.

63. Ibid., p. 262.

64. *Commonwealth* v. *Seay,* 1978 Mass. Adv. Sheets 2992 (December 4, 1978).

65. Rossman, "Impact," pp. 285–87.

66. Ibid., p. 312.

67. Ibid., p. 266.

68. Ibid., p. 78.

69. Ibid., p. 16.

70. Ibid., pp. 389–90.
71. Ibid., pp. 401–2.
72. Ibid., p. 406.
73. Ibid., p. 414.
74. Ibid., pp. 412–13.
75. Ibid., p. 416.
76. These figures were provided by Paul Froyd and were originally supplied by Frank Carney, of the Massachusetts Department of Correction, April 18, 1978. See also Michael Knight, "Boston Study Finds Gun Law Is Working," *New York Times*, June 18, 1979, sect. 4, p. 18.
77. Beha, "And Nobody," pp. 45–59.
78. Donald Newman, *Conviction: The Determination of Guilt or Innocence without Trial* (Boston: Little, Brown, 1966), p. 112.
79. See Milton Heumann and Colin Loftin, "Mandatory Sentencing and the Abolition of Plea Bargaining: The Michigan Felony Firearms Statute," paper presented at Conference on Plea Bargaining (French Lick, Indiana, June 1978). A shortened version by the same title was published in *Law and Society Review*, 13 (1979), pp. 393–430.
80. Ibid., p. 35.
81. Ibid., p. 34.
82. For an interesting discussion anticipating this general phenomenon, see Albert Alschuler, "Sentencing Reform and Prosecutorial Power: A Critique of Recent Proposals for 'Fixed' and 'Presumptive' Sentencing," *University of Pennsylvania Law Review* 126 (1978), pp. 550–83.
83. Ibid., p. 47.
84. In the preamble to the new code; see *Cal. Penal Code* 1170(a) (St. Paul, Minn.: West Publishing Co., 1980).
85. Sheldon L. Messinger and Phillip E. Johnson, "California's Determinate Sentencing Statute: History and Issues," in *Determinate Sentencing*, p. 16. My understanding of the history and operations of the California Adult Authority draws heavily on this article, other writings of Prof. Messinger, and discussions with him. See his "Strategies of Control" (Ph.D. dissertation, University of California, Los Angeles, 1969).
86. See George Jackson, *Soledad Brother* (New York: Coward-McCann, 1970).
87. See *In re Stanley*, 54 Ca. App. 3d 1030, 1033 (1976).
88. Caleb Foote, "Deceptive Determinate Sentencing," in *Determinate Sentencing*, p. 134.
89. Keith Hawkins, "Some Consequences of a Parole System for Prison Management," in D. J. West (ed.) *The Future of Parole* (London: Duckworth, 1972), p. 113.
90. Foote, "Deceptive Determinate Sentencing," p. 136. For a contrary opinion see Norval Morris, "Conceptual Overview and Commentary on the Movement Toward Determinancy," in *Determinate Sentencing*. See also Lynn Mather, *Plea Bargaining or Trial* (Lexington, Mass.: Lexington Books, 1979).
91. Interview with Sheldon Messinger, October 10, 1982.
92. See, for example, *In re Strum*, 11 Cal. 3d 258, 262 (1974).
93. *In re Lynch*, 8 Cal. 3d 410 (1972).
94. The court in *In re Wilkerson*, 271 Cal. App. 2d 798 (1969) stressed the concept that the Adult Authority had broad duties and responsibilities and exercised wide discretion in the administration of its powers. In *In re Schoengarth*, 66 Cal. 2d, 293, 302 (1967), the court held that "[o]ne who is legally convicted has no vested right

Notes

to the determination of his sentence at less than maximum." See also notes 95, 96, and 97 for developments in the courts' views.

95. Writings expressing these views are: *Task Force on Criminal Sentencing;* Von Hirsch, *Doing Justice;* and Marvin Frankel, *Criminal Sentences: Law Without Order* (New York: Hill and Wang, 1973).

96. See, for example, *In re Lynch,* 8 Cal. 3d 410 (1972), *In re Foss* 10 Cal. 3d 910 (1974), *In re Rodriguez,* 14 Cal. 3d 639 (1975), and *In re Stanley,* 54 Cal. App. 3d 1030 (1976). For a review of this history see Messinger and Johnson, "California's Determinate Sentencing," pp. 17–29.

97. *In re Rodriguez,* 14 Cal. 3d 639, 652 (1975).

98. *In re Stanley* and its companion case, *In re Reed,* 54 Cal. App. 3d 1030 (1976).

99. After a 1978 amendment that took effect on January 1, 1979, six classifications were added, all increasing possible sentences.

100. *Cal. Penal Code* 1170(b) (St. Paul: West Publishing Co., 1977) Supplement.

101. For a complete description of the law see Cal. Penal Code 1170(b) (St. Paul: West Publishing Co., 1977) Supplement, *et. seq.*

102. The preamble to the new law states: "The Legislature finds and declares that the purpose of imprisonment is punishment." *Cal. Penal Code* 1170(a) (St. Paul: West Publishing Co., 1980).

103. For reports on the effects of the new law on plea bargaining, see Jonathan D. Casper, David Brereton, and David Neal, *The Implementation of the California Determinate Sentence Law* (Washington, D.C.: National Institute of Justice, August 1981), chap. 3; and Pamela Utz, *Determinate Sentencing in Two California Courts* (Berkeley, Ca.: Center for the Study of Law and Society, March 13, 1981), pp. 32–135.

104. Casper et al., *Implementation,* pp. 204–206, 215–219.

105. Ibid., pp. 120–155.

106. The principal investigator of this comprehensive study of the effects of the new determinate sentencing laws in California and Oregon is Prof. Sheldon Messinger of the Law School at the University of California at Berkeley. He graciously made available to me early drafts of the multivolume report he and the research team were completing at the time of my work. The information reported in my text draws extensively on sections of this report. I am also deeply indebted to Prof. Messinger for sharing his knowledge of the history on sentencing in California. The paper prepared by Utz, note 103 above, is part of this study. The findings on county-based variations both before and after the new law are examined at length in one part of this voluminous report. See Richard Sparks, *Sentencing Before and After the DSL: Some Statistical Findings* (Newark, N.J.: School of Criminal Justice, Rutgers University, 1981).

107. Sparks, *Sentencing,* p. 34.

108. Ibid., pp. 37–67.

109. Ibid., p. 69.

110. *Sentencing Practices Under the Determinate Sentencing Law* (Sacramento, Ca.: Management Information Section, Board of Prison Terms, January 20, 1982), pp. 5–6. This report indicates that as of January 8, 1982, "the Board had found disparate sentences in 49 cases: 31 disparately high sentences and 18 disparately low sentences," and that the Board's motion to adjust sentences was "granted as to 14 defendants and denied to as to 17 defendants." In light of the variation found in the others' studies, these small numbers indicate that the Board is likely to act only in extreme situations. See note 106.

111. This is a variation of the well-known Hawthorne effect. For a good discussion of the independent impact publicity has on new laws, see Donald T. Campbell and

H. Laurence Ross, "The Connecticut Crackdown on Speeding: Time-Series Data in Quasi-Experimental Analysis," *Law and Society Review* 3: (1968), pp. 33–54.

112. The first of these reports to be made available is the excellent and careful three-county study by Jonathan Casper and his colleagues at Stanford University, Casper et al., *Implementation.* Another much larger comparative study of the California law, as well as determinate sentences in other states, is the Berkeley-based effort directed by Sheldon Messinger. See notes 104 and 106. See also Albert J. Lipson and Mark A. Peterson, *California Justice under Determinate Sentencing: A Review and Agenda for Research* (Santa Monica, Calif.: The Rand Corporation, 1980).

113. See, for example, Johannes Andenaes, "Deterrence and Specific Offenses," *University of Chicago Law Review* 38 (1971): 537–53. For a thorough review of the literature on deterrence and incapacitation, see Alfred Blumstein, ed., *Deterrence and Incapacitation* (Washington, D.C.: National Academy of Sciences, 1978), Franklin E. Zimring, "Policy Experiments in General Deterrence: 1970–1975," in Blumstein, *Deterrence*, p. 142.

114. William Chambliss, "Types of Deviance and the Effectiveness of Legal Sanctions," *Wisconsin Law Review* (1967): 703–19. For suggestive evidence of a widespread deterrent effect, see S. R. Barlow and M. R. Lomon, *White Collar Crime Project Evaluation* (Olympia, Wash.: Law and Justice Planning Division, 1976); M. K. Block, F. C. Nold, and J. G. Sidak, *The Deterrent Effect of Anti-trust Enforcement* (Stanford, Calif.: Center for Econometric Studies of the Justice System, September 1978); and E. Stotland, M. Brintnall, A. L'Heureux, and E. Ashmore, "Do Convictions Deter Home Repair Fraud?" in *White Collar Crime: Theory and Research*, ed., Gilbert Geis and Ezra Stotland (Beverly Hills: Sage Publications, 1980).

115. Much of the evidence for this is indirect, but strong. See, for example, David Sudnow, "Normal Crime: Sociological Features of the Penal Code in a Public Defender Office," *Social Problems* 12 (1965): 255–76; James Eisenstein and Herbert Jacob, *Felony Justice* (Boston: Little, Brown, 1977); Pamela Utz, *Settling the Facts* (Lexington, Mass.: Lexington Books, 1978); and Milton Heumann, *Plea Bargaining* (Chicago: University of Chicago Press, 1978). For a review and assessment of the research on sentencing and the issues of disparity and discrimination in sentencing, see the report of the panel of the National Academy of Science, Alfred Blumstein, ed., *Research on Sentencing: The Search for Reform* (Washington, D.C.: National Research Council, forthcoming). See also John Hogarth, *Sentencing As a Human Process* (Toronto: University of Toronto Press, 1971).

116. See Charles Silberman, *Criminal Violence, Criminal Justice* (New York: Random House, 1978), pp. 285–96.

117. See, for example, Vera Institute of Justice, *Felony Arrests: Their Prosecution and Disposition in New York City's Courts*, revised edition (New York: Longman, 1981); and Floyd Feeney, Forrest Dill, and Adrienne Wier, *The Deterioration of Criminal Cases* (Davis, Calif.: Center for Criminal Justice, 1981). More generally for a discussion of the value of "responsive law," see Philippe Nonet and Philip Selznick, *Law and Society in Transition* (New York: Harper and Row, 1977).

118. For a valuable survey of the development of the English law on appellate review of sentences, see D. A. Thomas, *Principles of Sentencing,* 2d ed. (London: Heinemann, 1979); and *Constraints on Judgment: The Search for Structural Discretion in Sentencing, 1860–1910* (Cambridge: Institute of Criminology, University of Cambridge, 1979). For a brief but interesting discussion of the current American concerns in light of the English experience, see idem,"Equity in Sentencing" (Sixth Annual Pinkerton Lecture, School of Criminal Justice, State University of New York, Albany, April 1977). See also idem, "Appellate Review of Sentencing and the Develop-

Notes

ment of Sentencing Policy: The English Experience," *Alabama Law Review* 20 (1968): 193–226. Thomas concludes: "A court exercising appellate jurisdiction over sentences can develop a meaningful case law of sentencing, *provided that it be prepared to take a sufficiently broad view of its functions and discard the normal approach of an appellate court in seeking only errors and abuses"* (emphasis added), p. 225.

See *ABA Project on Minimum Standards for Criminal Justice, Standards Relating to Appellate Review of Sentences,* Approved Draft, 1968 (New York: Institute of Judicial Administration, 1968).

119. I do not mean to suggest that adjudication is a cure-all. There are certainly limits to its value and effectiveness as a form of problem solving. But within the realm of what is within its grasp, adjudication can be a marvelously effective tool for ordering social relations, even complex ones. See, for example, Fuller, "The Forms and Limits of Adjudication," and Nonet and Selznick, *Law and Society in Transition.*

Chapter 5: Speedy Trial Rules and the Problems of Delay

1. *United States* v. *Ewell,* 283 U.S. 116 (1966), 120.
2. See Gregory P. N. Joseph, "Speedy Trial Rights in Application," *Fordham Law Review* 98 (April 1980): 611.
3. *Barker* v. *Wingo,* 407 U.S. 516 (1973).
4. Ibid., p. 523.
5. One commentator has suggested that only by "deconstitutionalizing" the right to speedy trial can it be implemented. See Joseph, "Speedy Trial," pp. 643–48.
6. Jack Weinstein, *Reform of Court Rule-Making Procedures* (Columbus: Ohio State University Press, 1977). See also Russell Wheeler, "Broadening Participation in the Courts through Rule-Making and Administration," *Judicature* 112 (1979): 280–90.
7. *Hearings before the Subcommittee on Constitutional Rights of the Committee on the Judiciary,* United States Senate, 92nd Congress, 1st Session, S. 895 (Washington, D.C.; GPO, 1971), pp. 1–207. Hereafter referred to as *Senate Hearings, 1971.*
8. *Senate Hearings, 1971,* p. 96.
9. 406 U.S. 979, 999–1000.
10. *Hearings before the Subcommittee on Constitutional Rights of the Committee on the Judiciary,* United States Senate, 93rd Congress, 1st Session on S. 754 (Washington, D.C.; GPO, 1973). Hereafter referred to as *Senate Hearings, 1973.* Statement submitted by Andrew H. Cohn, p. 220.
11. Testimony by Roland Kirks, director of the Administrative Office of the United States Courts, reported in *Hearings before the Subcommittee on Crime of the Committee on the Judiciary House of Representatives,* 93rd Congress, 2nd Session, on S. 754, H.R. 7873, H.R. 207, H.R. 658, H.R. 687, H.R. 773, H.R. 4807, *Speedy Trial Act of 1974* (Washington, D.C.; GPO, 1974), pp. 410–11. Hereafter referred to as *House Hearings, 1974.*
12. Ibid., p. 261, testimony of Daniel J. Freed.
13. Ibid., p. 385, Peter W. Rodino, Jr.
14. *Senate Hearings, 1973,* pp. 227–39, 242–84.
15. Rule 48(b) of the Second Circuit Rules. See *Senate Hearings, 1971,* pp. 339–93.
16. Ibid. See also Project, "The Speedy Trial Act: An Empirical Study," *Fordham Law Review,* 48: 718–79, 719, n. 53.
17. See Keith Boyum, "A Perspective on Civil Delay in Trial Courts," *Justice System Journal* 5 (winter 1979), 170–186.

18. *House Hearings, 1974.*

19. See, for example, Kenneth Mann, "The Speedy Trial Act Planning Process," *Harvard Journal of Legislation* 17 (winter 1980), 54–97, and Testimony of Alfonso J. Zirpoli, *House Hearings, 1974,* p. 383.

20. See, for example, *U.S.* v. *Howard* (Crim. No. Y-77-0387 E. D., s, Nov. 7 [1977]).

21. Mann, "Speedy Trial," pp. 14–15.

22. Ibid., p. 15.

23. In 1979, the ten-day indictment-to-arraignment period was eliminated and merged with a seventy-day indictment-to-trial period. Other changes were also instituted to meet the defense bar's objections to the law. See *Senate Report No. 96–212, Speedy Trial Act Amendments Act of 1979,* 96th Congress, 1st Session, June 13, 1979 (Washington, D.C.; GPO 1979). Hereafter referred to as *Senate Report 96–212.*

24. For a complete description of the plan, see Title 18, *U.S.C.,* Section 3166.

25. *Senate Report 96–212,* pp. 17–20.

26. These periods stem from amendments passed June 1979. The original act contained a ten-day time limit from indictment to arraignment and a sixty-day arraignment-to-trial period.

27. *U.S.C.* 3162 (1) (2).

28. Southern District of New York, "Plan for Prompt Disposition of Criminal Cases: Final Plan Pursuant to Speedy Trial Act of 1974," S.D.N.Y., June 30, 1978, pp. 111–19.

29. 18 *U.S.C.* 3161 (h) 8 (C).

30. 18 *U.S.C.* 3161 (h) 8 (A) [emphasis added].

31. See 18 *U.S.C.* 3174.

32. 18 *U.S.C.* 3161 (3) (B) (ii) and 18 *U.S.C.* 3161 (8) (B) (iv).

33. The Senate Judiciary Committee in its report on the 1979 Amendment states that any construction of the act to expand a defendant waiver right is "contrary to legislative intent and subversive of its primary objective: protection of the societal interest in speedy disposition of criminal cases by preventing undue delay in bringing such cases to trial." See *Senate Report 96–212,* p. 29. For a legal analysis of waiver under the act, and accompanying case citations, see Project, "Speedy Trial Act," pp. 753–65.

34. 18 *U.S.C.* 3161 (h) (4).

35. Ibid., 3161 (h) (G)

36. Section 3161 (h) (B) (6).

37. In addition, the planning provisions call for general background data on the administration of criminal justice in the district, description of the systems and procedures used to achieve and study the act's implementation, and requests for statutory amendments, rule changes, and appropriations necessary to meet the act's requirements.

38. See 18 *U.S.C.* 3165 (b).

39. See Project, "Speedy Trial Act," p. 747.

40. Mann, "Speedy Trial," p. 46.

41. *Senate Report 96–212,* p. 20.

42. Mann, "Speedy Trial," pp. 24–27, 38–41, 46.

43. Ibid., pp. 25–26.

44. Ibid., p. 26.

45. *Senate Report 96–212,* p. 27.

46. Ibid., p. 128.

47. Ibid., p. 36.

48. Mann, "Speedy Trial," pp. 81–83.

Notes

49. Ibid., pp. 85–86.

50. Quoted in Ibid., p. 75.

51. Ibid., p. 85.

52. Ibid., p. 95.

53. See *Report on the Implementation of Title I and Title II of the Speedy Trial Act of 1974* (Washington, D.C.: Administrative Office of the United States Courts, September 30, 1976) and *Second Report on the Implementation of Title I and Title II of the Speedy Trial Act of 1974* (Washington, D.C.: Administrative Office of the United States Court, September 30, 1977).

54. Northern District of Illinois, *N.D. Illinois, Speedy Trial Plan* (1976). More generally, see Mann, "Speedy Trial," and Richard Frase, "The Speedy Trial Act of 1974," *University of Chicago Law Review* 43 (1976): 667–723.

55. Ibid., pp. 72–73.

56. Mann, "Speedy Trial," p. 95.

57. Ibid., pp. 95–97.

58. Robert L. Misner, "District Court Compliance with the Speedy Trial Act of 1974: The Ninth Circuit Experience" (Report on file, Arizona University Law School) sect. II, p. 41, n. 127.

59. *Senate Report 96–212*, p. 16.

60. For a more extensive treatment of innovations, see Mann, "Speedy Trial," pp. 81–88.

61. Committee on Federal Courts, *Report Evaluating the Implementation of the Speedy Trial Act on the S.D.N.Y.* (New York: Association of the Bar of the City of New York, June 13, 1978). Hereafter referred to as *Bar Report*.

62. Second Circuit, *Council Memorandum: U.S. Atty. S.D.N.Y.*, January 9, 1973.

63. The Supreme and Appellate courts have ruled that neither the Speedy Trial Act's time limits nor the constitutional right of a speedy trial attaches until the government initiates a prosecution through indictment or arrest. See, for example, *United States* v. *Lovasco*, 431 U.S. 783 (1977); *United States* v. *Hillegas*, 578 F. 2d 453 (2d Cir. 1978). For a broad discussion of the adequacy of defense under the act, see Project, "Speedy Trial Act," pp. 738–53.

64. See, for example, Richard S. Frase, "The Speedy Trial Act of 1974"; and Office for Improvements in the Administration of Justice, *Delays in the Processing of Criminal Cases Under the Speedy Trial Act of 1974* (Washington, D.C.: U.S. Department of Justice, March, 1979), p. 32. Hereafter referred to as *Justice Department Report*.

65. *Justice Department Report*, p. 37.

66. *Report 96–212*, p. 27. See also *Annual Report of the Director, Administrative Office of United States Courts, 1981*, preliminary edition (Washington, D.C.: G.P.O., 1981), 86.

67. Robert L. Misner, "District Courts," p. 16.

68. See Project, "Speedy Trial," pp. 746–53. One judge, opposed to the act, was quoted as saying, "The best way to get rid of a bad law is to enforce it strictly."

69. Ibid., p. 24.

70. Ibid., p. 27.

71. *Plan for Prompt Disposition of Criminal Cases* (Compliance Report), Federal District of the Eastern District of Louisiana, April 30, 1976, pp. v–vi.

72. Office for Improvement in the Administration of Justice, *Delays in the Processing of Criminal Cases under the Speedy Trial Act of 1974* (Washington, D.C.: U.S. Department of Justice, March, 1979), p. 41.

73. *Klopfer* v. *North Carolina*, 386 U.S. 213 (1967).

74. *Barker* v. *Wingo,* 407 *U.S.* 514 (1972).
75. Thomas Church, Jr., *Justice Delayed: The Pace of Litigation in Urban Trial Courts* (Williamsburg, Va.: National Center for State Courts, 1978), pp. 48–49, 78. See also Midwest Research Institute, *Speedy Trials: A Selected Bibliography and Comparative Analysis of State Speedy Trial Provisions* (Mimeo, Kansas City, Missouri, April 14, 1978), pp. 145–81.
76. Ibid.
77. American Bar Association, *Standards,* sect. 1.3.
78. Richard C. Howard, "Criminal Procedure: The Slow Death of Oklahoma's Speedy Trial Statute," *Oklahoma Law Review* 31 (1978): 436–46.
79. Colorado Revised Statutes 18–1–405 (1973).
80. *People* v. *Ward,* 85 Mich. App. 473 (1978); *People* v. *Holbrook,* 60 Mich. App. 628 (1975). Of course, the court would respond that it was interpreting a Michigan statute and not the United States Constitution.
81. *Commonwealth* v. *Mayfield,* 496 Pa. 212, 364 A.2d 1345 (1976).
82. *Philadelphia Inquirer,* May 10, 1976, sec. B, p. 1. Quoted in Eric Brossman, "The Pennsylvania Prompt Trial Rule: Is the Remedy Worse than the Disease," *Dickensen Law Review,* 131 (1977): 237–64, 254, n. 130.
83. Merna B. Marshall and Joseph H. Reiter, "Pennsylvania Rule 1100: A Trial Court Working with Rule 1100," *Villanova Law Review* 23 (1978): 289.
84. Ibid., p. 290.
85. These data were supplied by Professor Wallace Loh of the University of Washington School of Law and are presented here with his kind permission.
86. "The Impact of Speedy Trial Provisions: A Tentative Appraisal," *Columbia Journal of Law and Social Problems* 8 (1972): 394–95.
87. Church, *Justice Delayed,* pp. 47–49.
88. Ibid., p. 61.
89. See Martin Levin, *Urban Politics and the Criminal Court* (Chicago: University of Chicago Press, 1977), pp. 226–45. See also Raymond Nimmer, *The Nature of System Change* (Chicago: American Bar Foundation, 1978), pp. 71–93, 143–56; and *Felony Arrests: Their Prosecution and Disposition in New York City's Courts,* rev. ed. (New York: Vera Institute of Justice and Longman, 1981).
90. David W. Neubauer, Marcia J. Lipetz, Mary Lee Luskin, and John Paul Ryan, *Managing the Pace of Justice* (Washington, D.C.: National Institute of Justice, September 1981), pp. 430–431.
91. See Nimmer, *System Change,* pp. 71–93; and Church, *Justice Delayed,* pp. 27–29.
92. Nimmer, *System Change*; Church, *Justice Delayed*; and Levin, *Urban Politics.*
93. Levin, *Urban Politics.*
94. Nimmer, *System Change,* pp. 143–56; Jane Hausner and Michael Seidel, *An Analysis of Case Processing Time in the District of Columbia* (Review Draft) INSLAW Research Project Publication No. 15 (Washington, D.C.: Institute for Law and Social Research, March 30, 1978).
95. Nimmer, *System Change,* pp. 143–56.
96. Levin, *Urban Politics,* p. 239.
97. Nimmer, *System Change,* pp. 143–56.
98. Laura Banfield and C. D. Anderson, "Continuances in Cook County Criminal Courts," *University of Chicago Law Review* 35 (1968): 256–316.
99. Macklin Fleming, "The Law's Delay: The Dragon Slain Friday Breathes Fire Again Monday," *Public Interest* 32 (1973): 13–33.
100. Neubauer et al., *Managing the Pace of Justice,* pp. 414–432.

Notes

Part 3 Assessment

Chapter 6: Impediments to Change

1. Raymond T. Nimmer, *The Nature of System Change* (Chicago: American Bar Foundation, 1978) pp. 176–77.
2. See, for example, David Rothman and Stanton Wheeler, eds., *History and Social Policy* (New York: Academic Press, 1981). This volume contains a number of articles by historians and policy makers, which convincingly demonstrate the value or the "lessons of history" for policy makers.
3. See, for example, Task Force Report on the LEAA of The Twentieth Century Fund, *Law Enforcement: The Federal Role* (New York: McGraw Hill, 1976); Samuel Krislov and Susan White, eds., *Understanding Crime* (Washington, D.C.: National Academy of Sciences, 1977); Malcolm M. Feeley and Austin Sarat, *The Policy Dilemma: The Crisis of Theory and Practice in the Law Enforcement Assistance Administration* (Minneapolis: University of Minnesota Press, 1980).
4. See Stuart Scheingold, *The Politics of Rights* (New Haven: Yale University Press, 1974).
5. See, for example, Murray Edelman, *The Symbolic Uses of Politics* (Urbana: University of Illinois Press, 1964); and *Politics As Symbolic Action* (Chicago: Markham Publishing Co., 1971).
6. Edmond Cahn, *Confronting Injustice* (Boston: Little, Brown, 1966)
7. Donald Horowitz, *The Courts and Social Policy* (Washington, D.C.: Brookings Institution, 1977)
8. See, for example, Jeffrey Pressman and Aaron Wildavsky, *Implementation* (Berkeley: University of California Press, 1973); Feeley and Sarat, *Policy Dilemma.*
9. Pressman and Wildavsky, *Implementation,* p. 109.
10. For a brilliant examination of policy articulation see Stuart Scheingold, *The Politics of Rights* (New Haven: Yale University Press, 1974); see also Murray Edelman, *Political Language: Words that Succeed and Policies that Fail* (New York: Academic Press, 1977).
11. David Rossman, Paul Froyd, Glen Pierce, John McDevitt, and William J. Bowers, *The Impact of the Mandatory Gun Law in Massachusetts* (Boston: Boston University Center for Criminal Justice, 1979).
12. For a fuller examination of the problems of criminal justice evaluation and evaluators, see Malcolm M. Feeley, "The Impact of Innovation: Assessing the Evaluations of Courts," in Don Gottfredson, ed., *Review of Criminal Justice Evaluation Results* (Washington, D.C.: National Criminal Justice Reference Service, 1979); see also Charles E. Lindblom and David K. Cohen, *Usable Knowledge: Social Science and Social Problem Solving* (New Haven: Yale University Press, 1979); Henry Aaron, *Politics and the Professors: The Great Society in Perspective* (Washington, D.C.: Brookings Institution, 1978).
13. Pressman and Wildavsky, *Implementation,* pp. 87–124.

Chapter 7: Toward a Strategy for Change

1. For elaboration on this perspective, see Herman Goldstein, "Improving Policing: A Problem-Oriented Approach," *Crime and Delinquency* 25, April 1979; 236–58.

For an important view emphasizing the need to understand the work of public service officials from a "bottom-up" perspective, see Michael Lipsky, *Street Level Bureaucracy,* (New York, Russell Sage Foundation, 1980). See also Norval Morris, "Impediment to Penal Reform," *University of Chicago Law Review* 33 (1966): 629–53. More generally, on the concept of responsive law, see Philippe Nonet and Philip Selznick, *Law and Society in Transition* (New York: Harper & Row, 1977).

2. See, for example, Yankelovich, Skelly, and White, Inc., "Highlights of a National Survey of the General Public, Judges, Lawyers, and Community Leaders," pp. 5–69; Barry Mahoney, Austin Sarat, and Stephen Weller, "Courts and the Public: Some Reflections on Data from a National Survey," pp. 83–102; and David Adamany, "The Implementation of Court Improvements," pp. 247–67; in Theodore Fetter, ed., *State Courts: A Blueprint for the Future* (Williamsburg, Va.: National Center for State Courts, 1978).

3. 347 U.S. 483 (1954); and 349 U.S. 294 (1955). 372 U.S. 335 (1963).

4. By far the best of this literature is Donald Horowitz, *The Courts and Public Policy* (Washington, D.C.: Brookings Institution, 1977). See also Raoul Berger, *Government by Judiciary* (Cambridge, Mass.: Harvard University Press, 1977); Macklin Fleming, *The Price of Perfect Justice* (New York: Basic Books, 1974); Nathan Glazer, "Towards an Imperial Judiciary?" *The Public Interest,* 41 (1975): 104–16; and Bayliss Manning, "Hyperlexis: Our National Disease," *Northwestern University Law Review* 71 (1977): 767–82.

But see Stuart Scheingold, *The Politics of Rights* (New Haven: Yale University Press, 1974); Abram Chayes, "The Role of the Judge in Public Law Litigation," *Harvard Law Review* 89 (1976): 1281–1316; Joel Handler, *Social Movements and the Legal System* (New York: Academic Press, 1977); and Ralph Cavanaugh and Austin Sarat, "Thinking About Courts: Toward and Beyond a Jurisprudence of Judicial Competence," *Law and Society Review* 14 (1978): 371–420.

5. Horowitz, *The Courts.* See also Lino Graglia, *Disaster by Decree: The Supreme Court Decisions on Race and the Schools* (Ithaca: Cornell University Press, 1976).

6. The classic statement on the need for new forms of rights in the modern welfare state is Charles A. Reich, "The New Property," *Yale Law Journal* 73 (1964): 733–87.

7. See Cavanaugh and Sarat, "Thinking About Courts."

8. James B. Jacobs, "The Prisoners' Rights Movement and Its Impacts," in Norval Morris and Michael Tonry, eds., *Crime and Justice,* Vol. 2 (Chicago: University of Chicago Press, 1980), pp. 429–70.

9. *Van Atta* v. *Scott,* Cal 613 p. 2d 210 (1980).

10. Horowitz, *The Courts.*

11. See, for example, Stuart Scheingold, *Politics of Rights.*

12. Ibid.; Handler, *Social Movements,* and Jacobs, "The Prisoners' Rights Movement."

13. For an account of the Los Angeles experience, see Brenda Hart Bohne, "The Public Defender as Policy Maker," *Judicature* 63 (1978): 176–84, 178.

14. For a discussion of the problems of the federal crime policy, see Victor Navasky, *Law Enforcement: The Federal Role* (New York: Twentieth Century Fund, 1976), and Malcolm M. Feeley and Austin D. Sarat, *The Policy Dilemma: The Crisis of Theory and Practice in the Law Enforcement Assistance Administration,* (Minneapolis: University of Minnesota Press, 1980).

15. Scheingold, *Politics of Rights.*

Index

Index

District courts, 131–32, 134, 160, 162–63, 170–73, 213
District of Columbia, *see* Washington, D.C.
Drug laws, 118, 136–37; *see also* Rockefeller Drug Law
Due process, 11, 13–15, 77n, 141, 215
Due Process Model, 17

Eighth Amendment, *see* Constitutional Amendments, eighth
England, 41, 114–15; courts, 24n, 32, 155, 212, 215
Ervin, Sam, 159, 164, 165, 209
Europe, 14, 27
Evidence, 5, 13, 32, 148

Failure-to-appear (FTA) rates, 44, 46, 48, 51, 54, 58–59, 63, 69, 71, 74, 75
Federal courts, 47, 74, 158, 161, 173, 211, 213, 222; *see also* Chicago, Illinois, federal courts; New York City, federal courts
Federal Judicial Conference, 159, 160, 163, 170
Federal Rules of Criminal Procedure, Rule 50 (b), 160–63, 164, 202, 204
Federal Speedy Trial Act, 9, 156, 163–77, 187, 201
Felonies, 28–29, 32, 42, 61, 65, 87, 102, 107, 132, 143, 145, 146; *see also* Assault; Robberies
Felony Firearms Statute, 9, 118, 137–38, 198, 202
Firearms, 5, 130; *see also* Bartley-Fox Gun Law; Felony Firearms Statute
Fleming, Macklin, 184–85
Flemming, Roy, 66–67, 68, 71
Florida, 152
Fogel, David, 115
Foote, Caleb, 43–44, 48, 77, 206, 209
Ford Foundation, 46, 51, 61, 62, 112
Foundation for Research and Community Development, 93
Fox, J. John, 129, 130

Frank, Jerome, 11
Frankfurter, Felix, 42
Freed, Daniel J., 46, 48, 66, 162, 186, 209
Friedman, Lawrence, 21
FTA rates, *see* Failure-to-appear rates

Gideon v. *Wainwright*, 211
Guilty pleas, 14, 20, 21, 22, 23, 95, 126, 132
Gun laws, *see* Bartley-Fox Gun Law; Felony Firearms Statute

Hawthorne effect, 200
Hogarth, John, 16–17
Horowitz, Donald, 197
Houston, Texas, 181, 182
HRA, *see* New York City, Office of Human Resources Administration

Illinois, 47, 172
Imprisonment, 16, 115, 122, 139, 140, 143, 144–45, 146, 151–52
Innovative programs, 7–8, 35–39, 69, 151, 209
In re Stanley, 143

Jacobs, James, 212
Jails, 41, 51, 183; overcrowding, 4–5, 52, 74
Judges, 4, 22, 54, 74, 103, 120, 159, 211; attitudes, 124, 163, 164–65; authority, 104–5, 133–34, 164–65, 198; role, 12, 18, 163; sentencing by, 116, 122, 126–27, 130, 138, 139, 144, 153, 154
Judicial discretion, 8, 13–17, 95, 129, 147–48, 150, 154, 210; *see also* Sentences, indeterminate

247

Index

Index

Index